ENVIRONMENTAL HEALTH –
PHYSICAL, CHEMICAL AND BIOLOGICAL FACTORS

UNINTENDED EFFECTS OF EMISSIONS REDUCTION

COMPETITIVENESS AND CARBON LEAKAGE ISSUES

Environmental Health – Physical, Chemical and Biological Factors

Additional books in this series can be found on Nova's website under the Series tab.

Additional e-books in this series can be found on Nova's website under the e-book tab.

ENVIRONMENTAL HEALTH –
PHYSICAL, CHEMICAL AND BIOLOGICAL FACTORS

UNINTENDED EFFECTS OF EMISSIONS REDUCTION

COMPETITIVENESS AND CARBON LEAKAGE ISSUES

KASSY KINSLEY
EDITOR

nova publishers
New York

For permission to use material from this book please contact us:
Telephone 631-231-7269; Fax 631-231-8175
Web Site: http://www.novapublishers.com

Library of Congress Cataloging-in-Publication Data

ISBN: 978-1-63117-902-0

Published by Nova Science Publishers, Inc. † New York

CONTENTS

Preface		**vii**
Chapter 1	Border Adjustments for Economywide Policies That Impose a Price on Greenhouse Gas Emissions *Bruce Arnold and Kevin Perese*	**1**
Chapter 2	International Trade and Carbon Leakage *Bruce Arnold*	**39**
Chapter 3	The Effects of H.R. 2454 on International Competitiveness and Emission Leakage in Energy-Intensive Trade-Exposed Industries *U. S. Environmental Protection Agency*	**65**
Index		**143**

PREFACE

Human activities around the world are producing increasingly large quantities of greenhouse gases (GHGs), the most abundant of which is carbon dioxide (CO2). In recent years, concerns about the effects those emissions might have on the climate have prompted Congress, federal regulators, and others to consider policies to reduce them. This book examines the unintended effects on the competitiveness of U.S firms of an economywide policy, such as a carbon tax or a cap-and-trade program, that would reduce emissions by imposing a price on them. The book also assesses the use of border adjustments, such as import tariffs and export subsidies, and transition assistance to mitigate those unintended effects.

Chapter 1 – Border adjustments could reduce the loss of competitiveness and make the costs of U.S. producers more similar to those of producers in countries that do not impose comparable policies, but such adjustments could be difficult to implement and to defend if challenged as being inconsistent with the General Agreement on Tariffs and Trade (GATT), one of the component agreements of the World Trade Organization (WTO). Transition assistance could also offset the loss of competitiveness and would probably be easier to implement but might engender difficulties under WTO agreements as well.

Chapter 2 – Under a broad-based carbon tax or cap-and-trade program, some of the reduction in U.S. carbon dioxide emissions would probably be offset by increases in foreign emissions that would not otherwise have occurred, a phenomenon known as carbon leakage. Industries with substantial total emissions, high trade ratios, and high emission intensities are the most likely to generate substantial leakage.

Therefore, the industries most likely to be sources of significant leakage through trade in their products are the chemical; primary metal (such as aluminum and iron and steel); and, to a lesser extent, the nonmetallic mineral products (cement, lime, gypsum, and glass) and petroleum and coal products (refining and coke production) industries. Under narrower programs targeting particular industries, significant leakage would occur in fewer industries.

Studies of economywide programs have produced estimates of leakage ranging from 1 percent to 23 percent of the emission reduction the programs would achieve in the countries implementing them. However, those estimates may not apply to future proposals, and estimating leakage is difficult and subject to considerable uncertainty.

Chapter 3 – Proposed greenhouse gas cap-and-trade legislation will play an important role in advancing both the United States' transition to a clean-energy economy and the long-run competitiveness of U.S. manufacturing. Some energy-intensive trade-exposed industries have raised questions, however, about the potential near-term international competitiveness impacts of domestic energy and climate legislation. If the adoption of a domestic cap-and-trade program leads some manufacturing activity and its associated emissions to shift to countries that do not yet have comparable greenhouse gas regulations, along with the economic concerns that this poses, this presents environmental concerns because the resulting "emission leakage" can undermine the environmental effectiveness of a domestic emissions cap. The most effective approach to address these concerns is to ensure significant action by all major emitters through ongoing international negotiations. It is important to understand, however, both the scope of potential competitiveness and leakage impacts that could exist prior to securing such action, and also the ability of certain provisions in proposed legislation — particularly the American Clean Energy and Security Act of 2009 (H.R. 2454) — to mitigate those impacts.

This report is a response to a request from Senators Bayh (D-Indiana), Specter (D-Pennsylvania), Stabenow (D-Michigan), McCaskill (D-Missouri), and Brown (D-Ohio). In a September 11, 2009 letter to the White House (see Appendix C), the Senators request "assistance in assessing the ... provisions [addressing energy-intensive trade-exposed industries] in Section 782(e) of ... H.R. 2454." In particular, they seek an analysis that determines which industrial sectors are likely to be eligible for allowance rebates to "trade-vulnerable" industries under H.R. 2454, and that assesses "potential competitiveness impacts to energy intensive and trade exposed manufacturers if the United States adopts a climate change policy...."

The Senators further request that the analysis consider the range of measures in H.R. 2454 that may benefit energy-intensive trade-exposed industries, and “identify additional data that, if provided, could improve the assessment of competitiveness impacts.”

In: Unintended Effects of Emissions Reduction ISBN: 978-1-63117-902-0
Editor: Kassy Kinsley

Chapter 1

BORDER ADJUSTMENTS FOR ECONOMYWIDE POLICIES THAT IMPOSE A PRICE ON GREENHOUSE GAS EMISSIONS*

Bruce Arnold and Kevin Perese

SUMMARY

Human activities around the world are producing increasingly large quantities of greenhouse gases (GHGs), the most abundant of which is carbon dioxide (CO_2). In recent years, concerns about the effects those emissions might have on the climate have prompted the Congress, federal regulators, and others to consider policies to reduce them. This Congressional Budget Office (CBO) report examines the unintended effects on the competitiveness of U.S firms of an economywide policy, such as a carbon tax or a cap-and-trade program, that would reduce emissions by imposing a price on them. The report also assesses the use of border adjustments, such as import tariffs and export subsidies, and transition assistance to mitigate those unintended effects. Border adjustments could reduce the loss of competitiveness and make the costs of U.S. producers more similar to those of producers in countries that do not impose comparable policies, but such adjustments could be difficult

* This is an edited, reformatted and augmented version of a report released by the Congressional Budget Office, dated December 2013.

to implement and to defend if challenged as being inconsistent with the General Agreement on Tariffs and Trade (GATT), one of the component agreements of the World Trade Organization (WTO). Transition assistance could also offset the loss of competitiveness and would probably be easier to implement but might engender difficulties under WTO agreements as well.

Policies That Impose a Price on Greenhouse Gas Emissions Could Have Unintended Effects on the Competitiveness of Some Industries

Under a carbon tax or cap-and-trade policy that covers all or almost all GHG emissions, industries' or firms' production costs would increase, causing those industries or firms to raise the prices of the goods they produce. The prices of emission-intensive goods (that is, goods whose production involves the emission of large amounts of greenhouse gases relative to their value) would increase the most. The higher prices would create incentives for other firms and consumers to purchase fewer of those goods, leading to declines in the production of those goods and the resulting emissions.

The higher prices would also make emission-intensive U.S. products less competitive relative to competing foreign products. Therefore, imports of the competing products would increase, and exports of the domestically produced products would decrease. Those changes in trade would lead to declines in the value of the dollar on foreign-exchange markets, thereby increasing the exports and reducing the imports of the products that are less emission-intensive. As a result, the policy would have little net effect on the U.S. trade balance, but it would have other unintended effects.

Foreign Emissions Would Increase

The displacement of domestically produced emission-intensive goods by products produced in other countries would cause some of the reduction in U.S. emissions to be offset by increases in foreign emissions. The resulting changes in trade in the products of the other industries would reduce foreign emissions, but not by as much as the changes in trade of emission-intensive products would increase them. Hence, the emission reduction achieved by the policy would be partially offset by a net increase in foreign emissions, a phenomenon known as *carbon leakage*. Such leakage would also occur through other mechanisms. In particular, reducing domestic demand for fossil fuels would lower the price of those fuels in other countries, thereby increasing their use in those countries.

The estimates in the literature that CBO reviewed for total leakage (through trade and other mechanisms) under an economywide carbon tax or cap-and-trade policies in the long term fall in a range from 1 percent to 23 percent of the reduction in emissions from domestic sources. Recently, a group of 12 modeling teams examined a policy consisting of a 20 percent reduction in emissions from their level in 2004. The modelers assumed that a large group of industrialized countries that includes the United States would implement the policy. The leakage rates projected by the modeling teams ranged from 5 percent to 19 percent, with a mean of 12 percent for the estimates. Leakage is likely to be lower in the shorter term, and studies of programs in Europe have found no significant leakage in the first few years of those programs.

Some Industries Would Lose Jobs and Profits

The displacement of domestically produced emission-intensive goods would also cause losses in employment and profits in the industries that produce them, such as the chemical and primary metals industries. The resulting changes in trade in the products of other industries would cause increases in employment and profits in those other industries. The initial net effect of all of the changes in trade on employment would probably be small, but some workers in emission-intensive industries who lost their jobs would have to look for work elsewhere.

Border Adjustments Could Offset the Loss of Competitiveness

Border adjustments are one way to offset the decline in competitiveness of U.S. emission-intensive firms and thereby reduce the consequent leakage and lost profits and employment that would otherwise result from an economywide carbon tax or cap-and-trade program. Those adjustments are provisions that would impose the same costs on imports that the emission-reduction policy would impose on domestic production and that would rebate the costs the policy would impose on domestic production that is exported. In the case of a carbon tax, the adjustments might take the form of import tariffs and export subsidies in proportion to the greenhouse gases emitted in the production of traded products. In the case of a cap-and-trade program, they might take the form of a requirement that importers obtain allowances for those emissions and that domestic manufacturers receive a rebate of the allowances required for their products that are exported.

Although border adjustments would help emission-intensive firms by preventing the increases in imports and declines in exports of their products that the policy would otherwise cause, they would harm firms that are not emission-intensive by preventing the declines in imports and increases in exports of *their* products that would otherwise occur as a result of changes in the exchange rate. Moreover, the adjustments would not reduce leakage occurring through mechanisms other than trade—in particular, leakage that would occur through lower prices for fossil fuels. Nevertheless, the 12-model study found that applying simplified border adjustments to emission-intensive industries with significant foreign competition reduced the mean estimated amount of leakage by roughly one-third.

Border Adjustments Would Be Difficult to Implement and to Defend if Challenged

Implementing border adjustments for an economywide carbon tax or cap-and-trade program would require determining the GHG emissions embodied in imports (that is, the amount of greenhouse gases emitted in producing each imported good and service), which would be extremely difficult if applied to most imports (but not if applied only to imports of fossil fuels). The difficulty could be reduced, but not eliminated, in several ways:

- By assuming that the emissions embodied in products were the same for all firms in a given industry and country;
- By limiting the adjustments to selected imports that embody large quantities of emissions or to primary products (such as steel and other metals) whose production would probably require lesser amounts of intermediate inputs purchased from other firms than would production of goods farther downstream (such as automobiles or other products made from steel); or
- By ignoring the emissions embodied in intermediate inputs.

However, such simplifications would result in adjustments that less accurately offset the effects of the emission-reduction policy on competitiveness.

The use of border adjustments might be challenged in the WTO as being inconsistent with the GATT, an agreement to which all WTO members, including the United States, are signatories. CBO has not independently analyzed the merits of such challenges but instead has reviewed several legal analyses of border adjustments in the literature. Those analyses suggest that the prospects for a successful defense of border adjustments are uncertain and would depend on the design of those adjustments. For example, an allowance requirement for imports under a cap-and-trade program might be more difficult to defend than would import tariffs and export subsidies under a carbon tax. Certain simplifications—such as assuming that the emissions embodied in products were the same for all firms in a given industry and country—might increase the difficulty of defending the adjustments in the WTO.

Transition Assistance Also Could Offset the Loss of Competitiveness

Transition assistance could also be used to reduce the decline in competitiveness that would result from a broad emission-reduction policy. Transition assistance would be the aid the government provided to compensate firms and their employees for some of the costs incurred while the emission-reduction policy was being phased in. If such assistance was made contingent on firms' levels of output or employment after the policy went into effect, it could offset the decline in competitiveness—and, consequently, some of the leakage and employment losses. In principle, such assistance could be made in an amount sufficient to eliminate the decline in competitiveness. However, doing so would negate the incentive for consumers to reduce their purchases of emission-intensive products—one of the major mechanisms of emission reduction under such policies. Hence, such assistance would result in more emissions under a carbon tax and would increase the cost of meeting the emissions cap in a cap-and-trade program.

Implementing transition assistance could be a substantial undertaking, but it would probably be easier than imposing border adjustments because the necessary data regarding firms' costs and emissions would be more readily available. However, it might engender problems under the Agreement on Subsidies and Countervailing Measures, a component agreement of the WTO.

EFFECTS ON FIRMS' COSTS OF POLICIES THAT IMPOSE A PRICE ON EMISSIONS

Human activities around the world are resulting in the emission of increasingly large quantities of greenhouse gases, the most abundant of which is carbon dioxide. Emissions can occur in a variety of ways. Emissions of CO_2 result directly from the combustion of fossil fuels, as when drivers burn gasoline in their cars or when utilities burn coal to produce electricity. However, many types of economic activity yield emissions indirectly, as when manufacturers, in the process of fabricating their products, consume electricity that was produced by a coal-fired generator or burn fuels for processes that require heat. Some products (such as metals and chemicals) require a great deal of energy to produce, while others (such as software) do not. Moreover, some energy-intensive products may be produced using electricity generated from nuclear power or renewable fuels, in which case no CO_2 emissions are generated. As a result, some products (such as crude oil or coal) directly contain carbon that will be emitted when the products are consumed, other traded products indirectly "contain" a great deal of carbon if their production is very energy-intensive and the energy used is largely derived from fossil fuels, and still other products are not very carbon-intensive at all.

In recent years, concerns about the effects that GHG emissions might have on the climate have prompted the Congress, federal regulators, and others to consider policies to reduce them. To date, the two main market-based approaches for reducing emissions that have received attention are a carbon tax and a cap-and-trade program:

- A *carbon tax* is a fee levied on the amount of carbon in fossil fuels or on the amount of CO_2 emissions that result when those fuels are burned.[1] Such a tax could be designed to include other greenhouse gases as well.
- Under a *cap-and-trade program*, the government sets gradually tightening limits on emissions, issues rights (or allowances) for firms to emit greenhouse gases consistent with those limits, and then lets firms trade the allowances among themselves.

A carbon tax would directly impose a price on emissions that would be subject to the tax. A cap-and-trade program would also effectively place a price on emissions; the price would be equal to the price of the emission allowances.

By placing a price on GHG emissions, the policies would create incentives for firms and consumers to reduce emissions in two main ways:

- Firms would be encouraged to reduce the quantities of greenhouse gases they emit during production, thereby lowering the amount they must pay for emissions. Similarly, consumers would be encouraged to reduce their emissions by, for example, setting thermostats lower in the winter and higher in the summer, driving less, and buying cars that are more fuel-efficient.
- Firms would probably raise their product prices to cover the cost imposed by the price on emissions. Consumers, in turn, would buy less of the products whose prices rose substantially because their production required substantial emissions. Instead, they would purchase goods whose prices rose less as a result of the policy because their production was less emission-intensive.

Policymakers have also considered and adopted other types of policies for reducing GHG emissions. Those policies include imposing a cap-and-trade program only for certain activities, such as generating electricity; adopting regulatory standards, such as more stringent standards for vehicle fuel economy; or providing subsidies for certain activities that reduce emissions, such as those for electricity produced from wind. Many of those policies also affect industries' competitiveness (see Box 1) although they are not the focus of this report.

UNINTENDED EFFECTS OF ECONOMYWIDE POLICIES THAT IMPOSE A PRICE ON EMISSIONS

Imposing an economywide carbon tax or cap-and-trade program would put the U.S. firms most affected— those that are emission-intensive—at a competitive disadvantage relative to their competitors in other countries unless those countries implemented similar policies. Such a policy would impose costs on domestic firms, allowing foreign producers from countries with less stringent policies, or no policy at all, to charge less for their goods than U.S. producers.[2] As a result, some consumers would switch to competing goods produced in other countries. That switching would unintentionally cause foreign production and thus foreign emissions to increase. Estimates of the increase in foreign emissions for economywide policies range from 1 percent to 23 percent of the reduction in domestic emissions achieved by the policies. Another unintended effect would be the loss of employment of workers in industries placed at a disadvantage by the policy.

The reduction in total employment from such a policy would probably be small; however, finding new work would take some time for the workers losing jobs.

Not all domestic producers would become less competitive under an economywide policy that reduced emissions by imposing a price on them. Some producers would benefit. That benefit would be large enough that the U.S. trade balance (that is, the total value of exports minus the total value of imports) would not be affected much, if at all, in the long term. The reason is that the policy would cause the value of the U.S. dollar to decline in foreign-exchange markets, effectively lowering the foreign-currency price of U.S. exports and raising the dollar price of U.S. imports.

The change in the value of the dollar would occur because the increase in imports of emission-intensive products prompted by the policy's effects on costs would require that dollars be used to purchase the foreign currencies needed to buy the products from abroad. Thus, all else being the same, the demand for foreign currencies would increase, driving up the dollar prices of the currencies (or, equivalently, driving down the foreign-currency prices of the dollar). Similarly, the reduction in exports would also drive down the price of the dollar.

The size of the effect on the value of the dollar would depend on the amount of the change in imports. Ultimately, the dollar's value would not decline enough to completely offset the higher costs paid by the producers most seriously affected by the emission-reduction policy—typically the producers that emit large quantities of greenhouse gases relative to the value of their products and cannot easily reduce their emissions.

However, the decline in the value of the dollar would more than offset the cost for the producers least affected—typically those that emit only small quantities or that could easily reduce their emissions—and would thereby give them a competitive advantage.

Consequences of the Changes in Competitiveness for Foreign Emissions

The switch to foreign emission-intensive products by U.S. and foreign consumers would cause foreign production and, consequently, foreign GHG emissions to increase—an example of *carbon leakage.*[3] The corresponding switch in demand from less-emissionintensive, foreign-produced goods to the competing U.S.-produced goods would reduce foreign production and thereby offset leakage through the trade of the more emission-intensive goods. However, the offset would only be partial because of the difference in emission intensity, resulting in a net increase in foreign emissions. That increase would offset to some extent the reduction in U.S. emissions achieved by the emission-reduction policy.

Leakage would also occur through mechanisms unrelated to the competitiveness of individual firms and industries. In particular, it would occur through changes in the prices of carbon-based fuels.[4] Increases in those fuel prices in the United States—and in the prices of goods requiring substantial amounts of such fuels to produce—would reduce U.S. consumption of those fuels. Because the fuels are internationally traded (primarily petroleum and, to a lesser extent, coal), the resulting reduction in demand would cause their prices in other countries to decline relative to what they would be otherwise.[5]

As prices dropped, demand for the fuels in other countries would rise and their emissions would increase. The amount of those changes would depend on the ease with which each fuel is traded and on other factors, such as whether foreign producers of fossil fuels—especially petroleum—react to changes in the United States. In particular, if Saudi Arabia or the Organization of Petroleum Exporting Countries lowered output to offset reduced consumption of oil in the United States—limiting the decline in the price of oil—leakage through lower fuel prices would be reduced.

The Location of Leakage

Most U.S. GHG emissions are not subject to leakage through trade. Roughly two-thirds of those emissions originate from the residential, commercial, construction, and transportation sectors, which (aside from international air travel) face no significant international competition.[6] (For that estimate, the emissions of the electricity-generating sector were attributed to each sector of the economy in proportion to its use of electricity.)

Emission reductions achieved in those sectors would not be subject to being offset by carbon leakage through trade in their products, although they would be subject to leakage through reductions in the international prices of carbon-based fuels.

Of the remaining one-third of U.S. emissions of greenhouse gases that occur in sectors where leakage is more likely to occur—agriculture, mining, and manufacturing—the amount of leakage that is likely to occur varies from industry to industry, depending on several factors. Particularly important are the following three:

- The industry's emissions;
- The amount of foreign competition, which can be roughly measured by the trade ratio (that is, the ratio of exports plus imports to U.S. market size); and
- The percentage increase in the industry's cost imposed by the policy, which is proportional to the industry's
- emission intensity—one measure of which is the ratio of its direct emissions to its value added.[7]

Despite the potential for leakage based on those and other factors, the benefits of agglomeration and the fixed location of existing plant and equipment, for example, tend to reduce the relocation of production to other countries, at least temporarily.[8]

Box 1. The Effects of Alternative Emission-Reduction Policies on Firms' Costs

Policymakers have considered and, in some instances, adopted policies other than a broad-based carbon tax or cap-and-trade program in order to reduce greenhouse gas (GHG) emissions. Such policies include price-based policies, such as a cap-and-trade program, for particular sectors of the economy; regulatory standards that are specifically designed to reduce GHG emissions; and subsidies for certain activities that result in lower emissions of greenhouse gases than do alternative activities. Each of those approaches would affect the economy or specific markets in different ways, and their impact on firms' costs would depend on how the policies were implemented.

Alternatives That Do Not Cover All of the Economy's Emissions

Rather than covering all or almost all emissions in the economy, a cap-and-trade program could be more narrowly targeted to cover the emissions of particular industries or sectors of the economy. In that case, prices would increase the most for the products of emission-intensive firms or plants in the targeted industries or sectors and for goods made by firms that use those products intensively. For instance, under such a policy that covered the electrical generating industry, the prices of electricity generated in an emission-intensive manner (such as from coal rather than from nuclear energy or hydropower) would increase the most along with the prices of goods whose production requires intensive use of such electricity. The subset of those firms that face foreign competition would see a decline in their competitiveness. Such a policy would probably result in lower leakage through the prices of fossil fuels than would an economywide policy that would also increase the prices of petroleum—not just the prices of coal and natural gas that would be affected by a policy limited to the electricity sector.

Alternatives That Impose Regulatory Standards

Another approach to reducing emissions would be to impose regulatory standards on certain industries or sectors. Such standards could regulate either production processes or products. Either kind of regulation would increase costs, leading to price increases and reduced sales of certain products—as would occur under a carbon tax or a cap-and-trade program. As in the case of a narrowly targeted cap-and-trade program, the largest price increases would be for products of the targeted industries and those of the industries that use those products intensively in their production; however, there would not necessarily be a direct relation between the cost and price increases and the emission intensity in the targeted industries or sectors.

Setting minimum requirements for the use of renewable fuels in electrical generating plants is an example of a *process standard*. A binding standard would increase costs by requiring the substitution of more expensive renewable fuels for the fuels the plants would otherwise use, thereby putting upward pressure on the average price of electricity and on the prices of the products of electricity-intensive industries.[1]

By increasing the average price of electricity, a renewable-fuels standard would raise the costs— and reduce the competitiveness—of firms whose production processes are electricity-intensive. However, the costs imposed on firms by a process standard would generally be different from those under an economywide carbon tax or cap-and-trade program.

Requirements that automobiles and household appliances meet specified energy-efficiency standards are examples of *product standards*. In particular, the existing fuel-economy standards for automobiles and other light-duty vehicles sold in the United States starting in 2017 require gradual reductions in GHG emissions.

Product standards would raise at least some firms' production costs and would probably lead to increases in the products' prices. In those instances, the rise in costs would not necessarily be proportional to the firm's emissions or the effect the policy would have on those emissions.

Moreover, unlike the case for process standards, the cost and price increases resulting from product standards would have little, if any, effect on the competitiveness of U.S.-made products—and, therefore, on trade—because the same standards would apply to a product sold in the United States regardless of whether it was produced domestically or imported from abroad.

Alternatives That Subsidize Low-Emissions Substitutes

Still another approach to reducing emissions would be to provide subsidies for alternative technologies that reduce GHG emissions. For instance, under current law, producers of electricity generated from wind power are allowed to take a tax credit for using that renewable source of energy.[2]

Such policies tend to reduce producers' costs rather than increase them, although not necessarily in proportion to the amount of greenhouse gases that are avoided. Those policies would result in lower prices for the products of firms receiving the subsidies and firms that intensively use the products of the firms receiving them. Therefore, they could increase the competitiveness of the most directly affected U.S.-made products.

[1] For additional information, see Congressional Budget Office, *The Effects of Renewable or Clean Electricity Standards* (July 2011), www.cbo.gov /publication /41451.

[2] For additional information, see the testimony of Terry M. Dinan, Senior Advisor, Congressional Budget Office, before the Subcommittee on Energy of the House Committee on Science, Space, and Technology, *Federal Financial Support for Fuels and Energy Technologies* (March 13, 2013), www.cbo.gov/publication /43993.

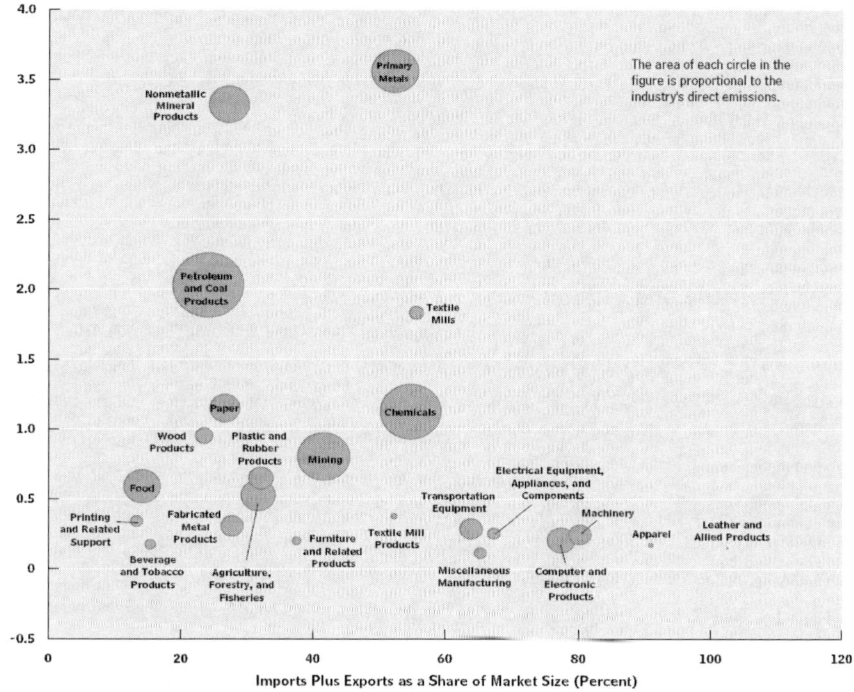

Source: Congressional Budget Office.

Notes: The largest amounts of leakage through trade are likely to occur in industries that have sizable emissions of carbon dioxide (CO_2), have high direct emission intensities, and face significant foreign competition as measured by the ratio of an industry's trade—imports plus exports—to the size of the U.S. market. The size of the U.S. market for each industry is measured here by the industry's output plus imports minus exports. As such, the ratio can exceed 100 percent.

Direct CO_2 emissions are the sum of an industry's own emissions and the emissions attributable to the production of the electricity it uses.

The CO_2 emission intensity shown is the direct intensity, which is the ratio of an industry's direct emissions to its value added (that is, to the value of its output minus the value of the intermediate inputs it uses), measured here in metric tons of CO_2 per thousand dollars of value added.

Figure 1. Carbon Dioxide Emission Intensities, International Trade, and Emissions, by Industry, 2010. (Carbon dioxide emission intensity).

Industries with substantial emissions, high trade ratios, and high emission intensities are the most likely to generate substantial leakage under an economywide carbon tax or cap-and-trade program that includes no provisions to reduce leakage (see Figure 1). The industries that stand out by those criteria are those that produce chemicals, primary metals (such as aluminum and iron and steel), and, to a lesser extent, nonmetallic mineral products (cement, lime, gypsum, and glass) and petroleum and coal products (refining and coke production). If such a policy were to reduce the non-CO_2 emissions of agriculture (particularly, methane and nitrous oxide)—something policies the Congress has considered to date would not do—and did so in a way that did not offset the cost with subsidies or other compensation, that sector would also probably be a sizable source of leakage.

The Magnitude of Leakage

Estimating leakage is difficult and subject to considerable uncertainty. Analyses of proposals for economywide policies have produced estimates of leakage rates ranging from 1 percent to 23 percent of the total emission reduction the policies would achieve in the countries implementing them.[9] Recently, a group of 12 modeling teams examined a policy consisting of a 20 percent reduction in emissions from their level in 2004 and assumed that it was implemented by industrialized countries listed in Annex I of the U.N. Framework Convention on Climate Change (including the United States but excluding the Russian Federation).[10] The leakage rates projected by the 12 teams ranged from 5 percent to 19 percent, with a mean of 12 percent for the 12 estimates. Industries that produce goods that are traded internationally are responsible for only a small portion of U.S. emissions (less than 37 percent of greenhouse gases and only 25 percent of CO_2). Hence, the loss of U.S. market share to foreign competitors and consequent percentage rates of leakage through trade could be much higher for some of those industries than for the economy as a whole if the estimates primarily reflect leakage through trade. However, the estimates of 1 percent to 23 percent also include leakage through the prices of carbon-based fuels. Estimates in the literature generally indicate that leakage through that mechanism may be more than half of total leakage, although such estimates are sensitive to parameters for which estimates are sparse and uncertain.[11]

Caution must be exercised in using those estimates to draw conclusions about the effects of various proposals, for several reasons. First, the estimates are generally for periods well into the life of a policy; leakage in the initial years would probably be much lower. Second, the reduction in competitiveness of U.S. producers and the consequent amount of leakage would depend on the number of U.S. trading partners that implemented comparable policies at the same time the United States did. Finally, the estimates generally do not account for provisions that could limit leakage by providing border adjustments or transition assistance linked to continuing levels of output or employment..

Some studies have looked back at the effects of policies already implemented in Europe and have found no evidence of significant leakage in the first few years of those policies.

One study, which looked at the effects in the United States of changes in energy prices from 1974 through 2001, concluded that some domestically produced goods were displaced by goods produced abroad in the first few years after prices rose.

However, the implied magnitude of leakage for an emission-reduction policy is not clear from the results. Regardless of the initial magnitude, leakage would probably increase after the first few years of a policy as transition assistance provided by the government to severely affected industries ended and firms had time to move their production abroad to avoid the higher costs.

Consequences of the Changes in Competitiveness for Employment

The declines in competitiveness of emission-intensive industries are of interest not only because of the resulting leakage but also because they would cause reductions in employment in some industries, forcing some workers to find work elsewhere. (Job losses would also arise from reduced total demand for the industries' products caused by consumers switching to other products—domestic or foreign—that are less emission-intensive.)[12] Similarly, the increase in competitiveness of other industries as a result of the decline in the value of the dollar would cause employment in those industries to rise. The initial effect on total employment—that is, the employment losses in industries placed at a disadvantage minus the employment gains in other industries—would probably be small, and even that small effect would subside as wages adjusted to equate the supply of labor with the demand for it. Nevertheless, some workers would be unemployed until they were able to find work in other industries or left the labor force.

Developing an estimate of the employment losses that each industry would incur as a result of the decline in competitiveness would require estimating the amounts by which exports would decrease and imports would increase, which is beyond the scope of this report. The industries most likely to lose large numbers of workers can be ascertained, however, by looking at the industries that have a high trade ratio and a high direct emission intensity and determining which of them has a large workforce.

When only CO_2 emissions—the main greenhouse gas for most industries and the only one for which CBO had emissions data for every industry—are considered, the largest employer in the sectors with a total trade ratio of at least 50 percent and a direct emission intensity of at least 1 metric ton of CO_2 per thousand dollars of value added is the chemical industry (see Figure 2). That industry employs about 1.2 million people.

The primary metals and textile mills industries are smaller than the chemical industry, employing about 450,000 people and about 150,000 people, respectively, but they have comparable trade ratios and higher direct emission intensities and therefore are more likely to lose market share to foreign rivals. The nonmetallic mineral products industry and the petroleum and coal products industry are also smaller, employing about 450,000 people and about 190,000 people, respectively, and also have higher direct emission intensities. However, their trade ratios are lower, making the likelihood of their losing significant market share to foreign rivals a little less clear. In principle, a small loss in competitiveness in an industry with a large workforce could cause a greater loss in employment than would result from a larger loss in competitiveness in an industry with a small workforce. However, such determinations would require estimates of employment losses and therefore are beyond the scope of this report.

When one considers all greenhouse gases (not just CO_2), the largest employer in the agricultural, mining, and manufacturing sectors that also has a significant trade ratio and a high emission intensity is the agriculture, forestry, and fisheries industry, which employs almost 2.2 million people. Although its CO_2 emission intensity ranks moderately low among the industries in question, its GHG emission intensity is more than five times as large and therefore ranks much higher.

However, under policies that do not limit or place costs on agriculture's non-CO_2 emissions, including the cap-and-trade programs the Congress has considered to date, that industry is not likely to have a substantial loss of employment.

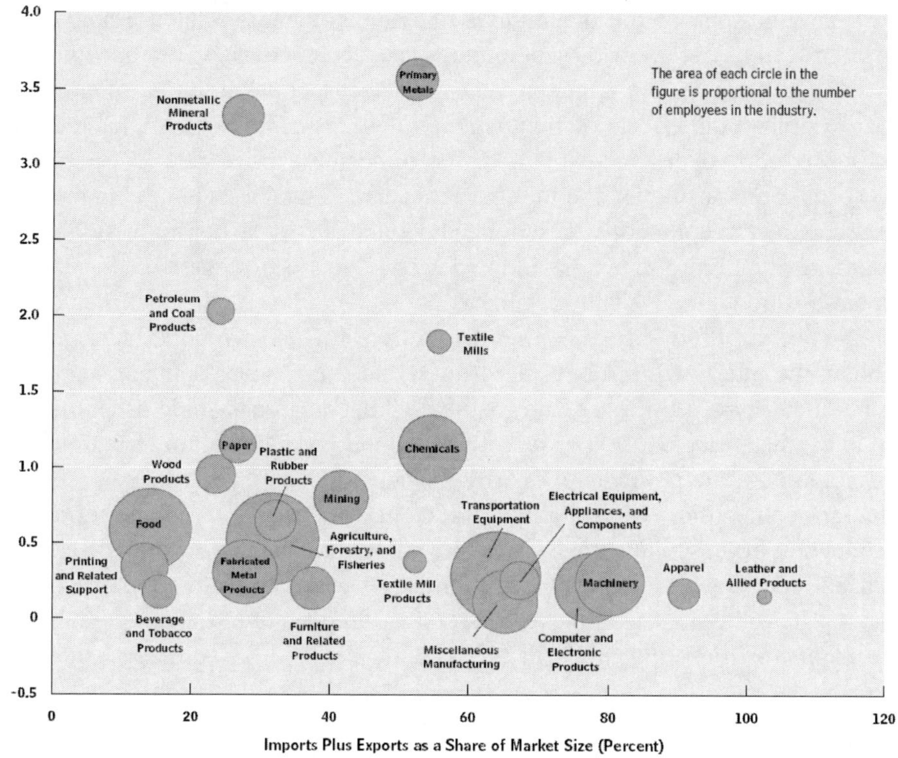

Source: Congressional Budget Office.

Notes: The largest employment losses resulting from reductions in international competitiveness are likely to occur in industries that have sizable employment, have high direct emission intensities for carbon dioxide (CO_2), and face significant foreign competition as measured by the ratio of an industry's trade—imports plus exports—to the size of the U.S. market. The size of the U.S. market for each industry is measured here by the industry's output plus imports minus exports. As such, the ratio can exceed 100 percent.

Direct CO_2 emissions are the sum of an industry's own emissions and the emissions attributable to the production of electricity that it uses.

The CO_2 emission intensity shown is the direct intensity, which is the ratio of an industry's direct emissions to its value added (that is, to the value of its output minus the value of the intermediate inputs it uses), measured here in metric tons of CO_2 per thousand dollars of value added.

Figure 2. Carbon Dioxide Emission Intensities, International Trade, and Employment, by Industry, 2010. (Carbon dioxide emission intensity).

A rough idea of the magnitudes of competitiveness-related employment declines can be ascertained by assuming that such declines would be similar in percentage terms to the competitiveness-related declines in output. One study that examined the effects of fluctuations in energy prices in the United States on net imports over time estimated that imposing a price of $15 per ton on CO_2 emissions would lead in the first few years to increases in net imports and consequently to competitiveness-related declines in domestic production of the products of emission-intensive industries equal to 1.0 percent to 1.3 percent of domestic production.[13] That decline is about one-third of the total decline in production in those industries that the study estimated would result from the policy; the other two-thirds results from reduced consumer demand stemming from higher prices. Another study estimated that a cap-and-trade program with an allowance price of $20 per ton of CO_2 would, in the medium term, increase net imports of the products of energy-intensive industries that face substantial foreign competition by roughly 1 percent to 2 percent of the domestic production of those industries.[14] Another study estimated that competitiveness-related reductions in output for such industries resulting from a tax of $15 per ton on CO_2 emissions would average roughly 46 percent of the total reductions in output for those industries in the medium run and 33 percent in the long run.[15]

Preserving Competitiveness with Border Adjustments

Border adjustments are one way to offset the decline in competitiveness of U.S. firms and thereby reduce the carbon leakage and lost profits and employment that would otherwise result from an economywide policy to reduce emissions by imposing a price on them.[16] The simplest border adjustments conceptually would be a tax (or tariff) on imports and a subsidy for exports that were proportional to the quantity of greenhouse gases emitted in the production of the good in question. The tax or subsidy might be lowered to reflect any increased costs resulting from an emission-reduction policy in the country from which the good was imported or to which it was exported. (Transition assistance linked to continuing levels of output or employment—a policy that is contained in some emission-reduction proposals to compensate firms and their employees for costs they would incur while the emission-reduction policy was phased in— could also be used to offset declines in competitiveness, as discussed below.)

Rather than impose a tax on imports, some proposals for a cap-and-trade program would require that importers obtain allowances for the emissions embodied in the goods they import. Such proposals would have separate pools of allowances for imported goods and those produced domestically, and the price of allowances for imported goods generally would be set at or below the market price of the allowances for domestic production and would serve the same function as a tax on imports. Eliminating the requirement for domestic production that is exported, or refunding the cost allowance for exported products, would provide a benefit to exporters of such products.

Border adjustments would prevent the increases in imports and declines in exports that would otherwise occur for emission-intensive goods. However, in doing so they would in turn prevent the decline in the value of the dollar and the resulting declines in imports and increases in exports of other goods. Because an emission-reduction policy would have little long-run effect on the overall trade balance, border adjustments to offset the policy's effects on trade would also have little effect on the overall trade balance. Similarly, the net effect of the adjustments on total employment would probably be small. If effectively implemented, however, the adjustments would reduce or eliminate the leakage that would otherwise occur through trade in emission-intensive goods— although not the leakage that would occur through lower prices of petroleum and other carbon-based fuels in other countries. The group of 12 modeling studies discussed above estimated leakage for the emission reduction program in question both with and without simplified border adjustments applied to the products of emission-intensive industries that face significant foreign competition. The mean estimated amount of leakage with the border adjustments applied was roughly one-third lower than the mean without the border adjustments.

The border adjustments could also hurt firms that are not emission-intensive—and firms that produce goods that are not traded, whether emission-intensive or not—for another reason: For a given goal for reducing domestic emissions, including border adjustments in an economy-wide carbon tax or cap-and-trade program would mean that less of the reduction would be achieved by industries that are emission-intensive and produce goods that are traded internationally, so more of the goal would have to be achieved by other industries.[17] Specifically, by reducing the loss of market share to foreign producers, the adjustments would increase the domestic output of goods in U.S. emission-intensive industries, and that increased output would increase

GHG emissions in the United States. Meeting the goal for reducing domestic emissions would therefore require greater reductions by other producers—those whose goods are less emission-intensive, are not traded, or both—than would be the case for a policy without border adjustments. In a cap-and-trade program, the increased output of emission-intensive industries would raise the demand for emission allowances, causing their price to rise. Those higher prices would increase the incentive for other industries to reduce their emissions.

PRACTICAL DIFFICULTIES IN IMPLEMENTING BORDER ADJUSTMENTS

In practice, implementing a policy to preserve the competitiveness of specific industries that would occur in the absence of the emission-reduction policy would be extremely difficult. Precisely offsetting the effects of the policy on competitiveness would require determining the amount of greenhouse gases that are embodied in imports and adjusting for existing policies to reduce GHG emissions in other countries. Modified versions of the adjustments could ease implementation somewhat but would still be difficult to carry out and would less accurately offset the policy's effects on the competitiveness of specific U.S. industries. Lowering adjustments for imports from countries that have their own emission-reduction policies would be consistent with the rationale for such adjustments but complicated to undertake.

Determining the Greenhouse Gas Emissions Embodied in Imports

Imposing border adjustments that would precisely offset all of the effects of a policy to reduce GHG emissions would require a lot of information that is not readily available. In particular, it would require knowing the total amount of greenhouse gases emitted in the production of each imported good, including the amounts emitted in the production of intermediate inputs used to make the imported good. Those amounts would not be evident from examining the imports, and determining the amounts would be extremely difficult and costly.[18]

Precisely Calculating the Ideal Border Adjustment

Precisely offsetting all of the effects that a policy that imposes a price on emissions has on competitiveness would require applying border adjustments to all imports and exports. In addition, the adjustment for a given product would have to be calculated separately for each firm (and perhaps for each individual plant) making the product, because the emission intensity of a good imported from one firm might differ from that of the same good imported from another firm. Such separate calculations would require gathering data on emissions from each firm. Moreover, they would require collecting data about the products each firm purchases for use as intermediate goods and from which other firms it purchases them, so that the emissions embodied in those goods could be calculated, which in turn would require collecting data from the companies producing the intermediate inputs about their emissions and purchases from other firms, and so on.

The task of gathering the required data and calculating the border adjustments might be given to a new entity similar to the one in the Department of Commerce's International Trade Administration (ITA) that, in conjunction with the International Trade Commission, enforces the antidumping and countervailing-duty (AD/CVD) laws.[19] When U.S. imports are suspected of being dumped or subsidized, the foreign producers of those goods are required to report to the ITA the information it needs to determine whether the imports are being dumped or subsidized and, if so, by how much. Using that information (and information obtained elsewhere), the ITA makes a determination. A similar arrangement might require producers of U.S. imports to report to the ITA or to another U.S. government agency the information needed to calculate the emissions embodied in those imports.

Two differences between enforcing the AD/CVD laws and implementing border adjustments for an emission-reduction policy could cause greater difficulties for the latter, however. First, investigations for AD/CVD law enforcement occur only for imports suspected of being dumped or subsidized, which constitute only a small fraction of all U.S. imports. By contrast, for border adjustments applied to all products, all firms producing U.S. imports and all significant suppliers of intermediate inputs to those firms would have to be investigated to determine the emissions embodied in U.S. imports—a much larger task.

Second, to create an incentive for firms under investigation to cooperate, U.S. AD/CVD laws stipulate that if a firm does not provide the necessary information within specified deadlines, the ITA may use "facts otherwise available" in its place. Such facts could be information supplied by the domestic U.S. industry that filed the AD/CVD case, which has an interest in having duties imposed on the imports with which it competes. Thus, the use of such facts may result in a dumping or subsidy determination that is larger than would be the case if it had been made on correct information supplied by the firm being investigated—which encourages the firm being investigated to provide such information. Provisions of the WTO's Antidumping and Subsidies Codes explicitly allow governments to collect data from firms for the enforcement of AD/CVD laws, and they explicitly allow the use of facts otherwise available when the firms being investigated do not provide the necessary information.[20] However, the WTO agreements do not contain similar language that would apply to collecting information for border adjustments for an emission-reduction policy.

Approximating the Ideal by Treating All Firms in Each Industry of Each Foreign Country the Same

The implementing authority could simplify its task by assuming that all imports of a given product from a given country embody the same quantity of emissions even if different units of the product were produced by different firms in different locations. Under that assumption, the emissions embodied in imports could be calculated using input-output tables for the economies of the countries producing the imports. Entries in such tables indicate the quantity of inputs that each industry purchases from each other industry.

Using that procedure would entail several problems, however. First, even the most detailed input-output tables available are likely to mix together industries that are distinct in ways that could yield an inaccurate calculation for the purpose at hand. For example, such tables generally lump together electricity produced from nuclear power, which produces no GHG emissions, and electricity produced from the burning of coal, which produces substantial quantities of such emissions. As a result, calculating the emissions of any industry that uses electricity would attribute average emissions per unit of electricity of the whole electricity-generating sector to that industry even if the electricity the industry purchased was produced by a nuclear power plant and therefore embodied no emissions.

Second, some of the intermediate inputs that the foreign producers of imports used in their production would have been purchased from other countries. Therefore, the calculation would require an input-output table not only for each foreign producer's country but also for each other country from which that producer purchased significant quantities of intermediate inputs. Using input-output tables from several countries for one calculation may be difficult because the tables are likely to divide their respective economies in different ways, with the industry definitions used in a given table overlapping only partially with those used in other tables.

Third, even in industrialized countries like the United States, input-output tables—especially detailed ones— take so long to produce that the most recent ones are likely to be five or more years out of date, possibly leading to inaccurate estimates for rapidly growing industries. And finally, the procedure might be successfully challenged in the WTO (as discussed below).

Making Adjustments Only for Selected Products

Another way to simplify the task of determining the emissions embodied in imports would be to impose border adjustments only on selected imports that embody large quantities of emissions.

Limiting the adjustments to primary products, such as steel and other metals, might also alleviate problems with making determinations about emissions because many of the firms producing such products would probably purchase fewer intermediate goods than would other firms, many of which use primary products (or products made from them) as intermediate inputs.

Making Adjustments Based on Direct Emissions Only

Still another simplification would be to base the adjustment for each import only on the greenhouse gases emitted directly by the firm that produced the imported good (and perhaps the emissions attributable to the electricity the firm used) and not on the emissions embodied in intermediate inputs. If the adjustment was based on the ratio of direct emissions to the value added by the firm—that is, the value of the product minus the value of intermediate inputs used to produce it—then the adjustment would be either slightly too high or slightly too low, depending on how that ratio compared with the same ratio for the intermediate inputs.

Such an adjustment would also make it easier for firms to take advantage of the system.

For example, a firm that previously produced both a final product and its intermediate inputs might rearrange its production so that the intermediate inputs were purchased from a nominally independent firm, whose emissions would not count under the simplified procedure. As a result, the firm would face lower border adjustments.

Modifying Border Adjustments to Reflect Other Countries' Emissions Policies

Provisions in many cap-and-trade proposals have specified that border adjustments be lowered for imports from countries that have their own emission-reduction policies. The adjustments might have to be lowered if they are to be successfully defended from a challenge in the WTO. However, a provision for lowering the adjustments could make them much harder to implement because determining the amount by which the border adjustments should be reduced for each country to reflect its own emission-reduction policy could be difficult. For imports from a country that has a carbon tax or a cap-and-trade program covering all industries, determining the appropriate adjustment would not be particularly difficult. It would involve simply subtracting the carbon tax rate or the price of emission allowances in that country from the border adjustments that the United States would otherwise make for imports from the country. However, determining the adjustment would be more difficult if the country's carbon tax or cap-and-trade program did not apply to all of its industries. Industries that were not directly covered by the emission-reduction policy might still have higher costs because they purchased inputs from the industries to which the program applies; hence, the United States could not simply reduce or eliminate its border adjustments for imports from the industries to which the foreign country's program applied and not do the same for other industries in that country. Instead, the United States would have to track the purchases of intermediate inputs by the various industries and thereby trace the amount by which the country's program raised those industries' costs. Making the determination would be even more difficult for imports from countries whose emission-reduction programs were neither a carbon tax nor a cap-and-trade program but instead were a program of regulations or standards that were not market based. In almost all such cases, making a reasonably accurate determination would not be possible.

Potential Legal Challenges to Border Adjustments under the GATT

One or more WTO member countries might view border adjustments as inconsistent with U.S. obligations under the GATT and might therefore challenge the adjustments in the WTO.[21]

One of the WTO agreements, the Dispute Settlement Understanding, lays out the procedures for settling trade disputes.[22] The aggrieved country must first consult with the country imposing the offending policy to attempt to resolve the dispute. If the consultations fail, it may then challenge the policy by requesting that the WTO convene a dispute settlement panel. After hearing the arguments by both parties, the panel issues a ruling, which may be appealed to the Appellate Body for a final decision. If the final ruling is in favor of the challenging country, the Appellate Body can grant remedies to that country, such as requiring the defending country to withdraw the offending policy or authorizing the challenging country to retaliate by imposing trade restraints.

A number of years might pass before rulings in challenges of various countries' implementations of border adjustments make clear whether border adjustments for emission-reduction programs are consistent with the GATT and which particulars of such adjustments are consistent with the GATT and which are not.

Given the possibility that border adjustments would be challenged, CBO reviewed several legal analyses of them in the literature.[23] The analyses are based in part on rulings that have been issued in previous disputes. Decisions by dispute settlement panels and the Appellate Body of the GATT and the WTO are made on a case-by-case basis and serve only as rough guides—not as binding law—for how a future dispute settlement panel or the Appellate Body would decide a similar dispute. Therefore, the literature's discussion of past decisions shows only the past treatment of similar GATT provisions; it does not necessarily predict future rulings of a dispute settlement panel or the Appellate Body. Analysis of the text of the WTO agreements and relevant jurisprudence is sufficiently complicated that an in-depth discussion of all of the issues that might arise is not possible in this report; the following discussion merely outlines some of the more likely issues. In addition, the discussion is based solely on the analyses that CBO reviewed and does not reflect any independent legal analysis by CBO.

On the basis of that review, it appears that border adjustments would probably violate a general provision of the GATT, so they would have to be defended using one of the exceptions to the general provisions. Whether border adjustments could be successfully defended is unclear, and the likelihood of a successful defense of them would depend on their design. For example, border adjustments for a cap-and-trade program could be more difficult to defend than those for a carbon tax. Simplifying implementation in certain ways, such as assuming that the emissions embodied in products were the same for all firms in a given industry and country, might make border adjustments more difficult to defend.

Conflicts with General Provisions of the GATT

Article II:1(b) of the GATT prohibits each member country from imposing import tariffs or other charges that are higher than upper limits (called tariff bindings or bounds) listed in a schedule agreed to by that country and annexed to the agreement.[24] Almost all U.S. tariffs are at or near their bindings, so border adjustments in the form of tariffs or other charges on imports would make the total of tariffs and other charges on many U.S. imports higher than the bindings the United States has agreed to and thus would violate Article II:1(b).

Cap-and-trade proposals, however, have typically included provisions for border adjustments based on emission allowances. Article XI:1 of the GATT proscribes prohibitions and quantitative restrictions on imports. A prohibition on imports for which the importer does not have emission allowances conceivably could be ruled as a violation of that provision. Alternatively, the requirement could be ruled as not violating that provision because the allowances could be purchased and the imports therefore are not actually prohibited. However, such reasoning might imply that the requirement is effectively a charge on the imports and therefore violates Article II:1(b).[25]

Possible Defenses

Whether adjustments took the form of tariffs or a requirement of allowances for imports, a general provision of the GATT would probably be violated, so border adjustments would have to be defended using an exception to the general provisions of the GATT. Two exceptions that could be useful in a defense are contained in Article II:2(a) and Article XX.

Article II:2(a) Defense

Notwithstanding the general prohibition in Article II:1(b) on tariffs and charges that are higher than the tariff bindings to which a country has agreed, Article II:2(a) provides that a country that imposes a tax internally on a domestically produced good can impose an equivalent tax on imports of "like" goods. That provision, for example, allows the European Union to impose its value-added tax on imports and to rebate it on exports.

Several issues would be likely to arise when asserting an Article II:2(a) defense of border adjustments for an emission-reduction policy. Foremost is whether the policy and the associated requirement of allowances for imports are "taxes" within the meaning of Article II:2(a). Strictly speaking, a cap-and-trade program and an associated requirement of allowances for imports are not taxes, but their effects are sufficiently similar to those of taxes that they conceivably could be treated as such. One analysis that CBO reviewed states that "the general definition of a tax is a compulsory contribution imposed by the government for which taxpayers receive nothing identifiable in return."[26] Although the requirement for emission allowances is not a contribution, it resembles one in the sense that firms must spend money to purchase them, which from the firms' standpoint is like a contribution, and the firms receive nothing in return for the expenditure. Hence, the analysis argues, the requirement might qualify as a tax.

Another issue that would be likely to arise in the case of a cap-and-trade program is whether the program is still a tax if the government distributes some or all of the emission allowances for imports free of charge. If a firm did not have to purchase its required allowances, the requirement would not resemble a compulsory contribution and, therefore, would not appear to be a tax. However, not all firms are likely to be given the allowances they need; some firms would have to purchase at least some of their allowances. Moreover, the effects of a cap-and-trade program on the output and employment decisions of firms are mostly independent of whether the government sells the emission allowances or gives them away. A firm that receives the allowances it needs free of charge incurs a cost in using them for its own production because it thereby forgoes the revenue it could receive from selling them. Thus, the analysis argues, a cap-and-trade program might still be considered a tax even if the government gave the allowances away.

The legal analyses CBO reviewed indicate that, assuming that a dispute settlement panel and the Appellate Body considered an emission-reduction policy and associated border adjustments to be taxes on carbon emissions, several criteria would need to be met to successfully assert an Article II:2(a) defense. Some of the issues the analyses address involve process versus product taxes, the so-called national treatment requirement, and the so-called most favored-nation requirement.

Is the tax on GHG emissions a process tax or a product tax? A tax on emissions would seem to be a tax on the production process rather than on the product itself, whereas the phrase "in respect of the like domestic product" in Article II:2(a) would seem to suggest that only product taxes are eligible for that provision. Some analysts interpret the relevant jurisprudence as suggesting that taxes on inputs incorporated into a product would be ruled eligible but that taxes on inputs like energy that are consumed during production but not incorporated into the product would probably be ruled ineligible. If so, then a cap-and-trade program would most likely be ruled ineligible for the Article II:2(a) provision allowing an equivalent tax on imports of "like" goods. Other analysts, however, argue that the eligibility of energy taxes for border adjustments is unclear, and they present reasons to think that such taxes might be ruled eligible.[27]

Do the policy and adjustments give national treatment to imports? That is, is the border tax on each imported good less than or equal to the emissions tax on the "like domestic product" as required by Article III:2 of the GATT?[28] The critical issue here is the meaning of like domestic product under WTO jurisprudence. In past cases, like products have been determined by their characteristics, and the processes used to produce them have been considered irrelevant. If that rule was used in assessing the border adjustments for a tax on GHG emissions, the adjustments might be found to violate the national treatment requirement in any case in which larger quantities of those gases were emitted in the production of an imported product than in the production of the like domestic product.[29] Such a finding would void a defense under Article II:2(a), which requires that the relevant tax be consistent with Article III.

Do the policy and adjustments adhere to the most-favored nation requirement of Article I? Article I requires that imports of each product from each WTO member be treated no less favorably than imports of like products from every other member, a requirement commonly referred to as the most-favored-nation requirement. Thus, a country cannot charge one tariff on a product imported from one WTO member and another tariff on a like product imported from a different member.[30] Once again, the meaning of "like product" is critical because two otherwise identical products would require different border adjustments if different amounts of greenhouse gases were emitted in their production. If those otherwise identical products were considered to be like products, then the most-favored-nation requirement would be violated, once again voiding an Article II:2(a) defense.

The requirement also would probably mean that border adjustments could not be reduced for imports from countries that had emission-reduction policies of their own. Countries that had their own carbon tax or cap-and-trade programs and disliked the terms of the U.S. policy would have to rebate their tax on, or the emission allowances required for the production of, the goods they export to the United States.

Article XX Defense

Article XX of the GATT is devoted to general exceptions to the provisions of the GATT. The most relevant exceptions for an emission-reduction policy with border adjustments are those contained in Article XX:(b) and Article XX:(g). The article's introductory paragraph (also called the *chapeau* or proviso), and the two paragraphs in question, follow:

> Subject to the requirement that such measures are not applied in a manner which would constitute a means of arbitrary or unjustifiable discrimination between countries where the same conditions prevail, or a disguised restriction on international trade, nothing in this Agreement shall be construed to prevent the adoption or enforcement by any contracting party of measures: . . .
>
> (b) necessary to protect human, animal or plant life or health; . . .
>
> (g) relating to the conservation of exhaustible natural resources if such measures are made effective in conjunction with restrictions on domestic production or consumption . . .

Several analyses that CBO reviewed note that "relating to" is a less stringent standard than "necessary to" and therefore is more easily met.[31] Hence, this discussion focuses on a defense based on Article XX:(g). According to the analyses that CBO reviewed, some issues a panel would be likely to consider in assessing such a defense are as follows:

- *Is the atmosphere an exhaustible natural resource?* Several analyses that CBO reviewed noted that clean air has previously been found to be an exhaustible natural resource.[32] One analysis says that for the United States to prevail, the dispute settlement panel would have to find that a "sufficient nexus" exists between CO_2 emissions of the country challenging the border adjustment and consequences for the climate in the United States.[33]
- *Are the border adjustments related to the goal of conserving that natural resource (that is, to mitigating warming of the atmosphere)?* According to one legal analysis, WTO panels have interpreted "relating to" to mean "primarily aimed at."[34] That analysis argues that whether border adjustments would satisfy that test is unclear because

estimates suggest that the leakage the adjustments address might be small even without them.

- *Does the adjustment comply with the requirement in the introductory paragraph of Article XX that the "measures are not applied in a manner which would constitute a means of arbitrary or unjustifiable discrimination between countries where the same conditions prevail, or a disguised restriction on international trade"?* According to one legal analysis that CBO reviewed, several prominent environmental cases in the WTO have turned on the introductory paragraph.[35] Similarly, another analysis says that in all cases for which an Article XX defense has failed before the Appellate Body, it failed because the policy at issue was ruled to be contrary to that paragraph.[36]

Several related issues could arise under the introductory paragraph.[37] First, the apparent purpose of the border adjustments would have to be a "rational connection" to the conservation goal and not an attempt to "level the playing field" to protect domestic industries.

Second, the United States would probably have to determine the border adjustments for each firm individually rather than for all firms from a given country as a group. Such detailed determinations would be very difficult to make, and simplifications to that task might be found to be unjustifiable. One analysis argues that lack of usable data on foreign emissions may raise questions about whether border adjustments can be applied in a nondiscriminatory fashion.[38]

Third, the United States could not require that another country have an emission-reduction policy that was the same as or similar to the U.S. policy, or one that used a market mechanism, as a condition for not imposing border adjustments on the imports from the country. It could only require that the other country's program be "comparable in effectiveness" to the U.S. policy.

Fourth, some legal analyses suggest that border adjustments might have to be reduced for imports from countries that have their own emission-reduction policies and perhaps for imports from some developing countries. Estimating the amount of such reductions would be difficult, especially for emission-reduction programs that consist of regulations or standards and are not market based.

Fifth, according to some analyses, WTO case law indicates that before imposing the adjustment the United States must have "serious, across-the-board negotiations" with countries that would be affected by the adjustments.

PRESERVING COMPETITIVENESS WITH TRANSITION ASSISTANCE

Instead of using border adjustments to preserve the competitiveness of U.S. firms, policymakers could consider providing transition assistance to reduce the decline in competitiveness that would result from a carbon tax or a cap-and-trade program. Transition assistance would be government aid provided to firms to reduce the costs of complying with an emission-reduction policy. If such aid was made contingent on the firms' levels of output or employment after the policy took effect, it would subsidize production and thereby tend to offset the effects that the policy would have on competitiveness and the consequent leakage and employment loss. It would also, however, cause a greater reduction in the effectiveness of the policy at reducing emissions than would be caused by border adjustments. Such assistance would probably be easier to implement than border adjustments because of differences in availability of the necessary data, but like border adjustments it might engender problems in the WTO.

Transition Costs and Transition Assistance

Even in the absence of trade, an emission-reduction policy would lead to a contraction of some industries or firms and consequent losses to their stockholders and employees. The costs could be high during the transition to the new lower levels of output and consumer demand for certain products. During that period, industries would incur the cost of retooling their factories to reduce their emissions, and their carbon tax payments or costs of purchasing emission allowances would be particularly high until the retooling was complete. Employees who were laid off would probably experience a period of unemployment as they searched for new employment.

To alleviate transition costs, some proposals for economy-wide policies that would impose a price on emissions have contained provisions for transition assistance to firms that would be severely affected. Such assistance might take the form of direct monetary payments or tax credits for retaining certain levels of output or employment (in the case of a carbon tax) or free emission allowances from the government in proportion to the emissions a firm had before the policy went into effect (in the case of a cap-and-trade program).

Implications for Competitiveness

If transition assistance was tied to the level of output or the number of workers that a firm continued to employ after an emission-reduction policy went into effect, it would offset the negative effects that the policy would have on the international competitiveness of U.S. emission-intensive industries by effectively subsidizing firms' output or employment. Such assistance would aid the stockholders and the employees of emission-intensive firms, and, in principle, could be designed to completely eliminate those negative effects. However, in doing so, it would also eliminate the incentive that the emission-reduction policy would otherwise create for consumers to reduce their purchases of emission-intensive products. Firms would still have an incentive to reduce the emission intensity of their production, but eliminating the incentive for consumers would make the emission-reduction policy less effective. As a result, less emission reduction would be achieved under a carbon tax, and the cost of meeting the emissions cap in a cap-and-trade program would be higher.

If the assistance was not made contingent on continued production or employment levels by the firm or industry, however, it would not affect firms' international competitiveness or help their employees; it would only ameliorate the effects of the policy on the stockholders of the firms receiving the assistance. Specifically, assistance that was not contingent on production or employment levels would not affect firms' decisions concerning how much to produce, how many workers to employ, or the prices to charge for their products. That would be the case for direct payments because such payments would not change the fact that raising the level of production would increase a firm's carbon tax bill. And that would be the case for free emission allowances because such allowances would not change the fact that raising the level of production would increase the revenue a firm would forgo by not being able to sell the corresponding allowances. Thus, firms would face the same cost trade-off in making their production decisions whether they received such transition assistance or not.

Practical Difficulties

The practical difficulties of providing transition assistance would probably be considerably less than those for border adjustments. The reason is that, unlike the case for border adjustments, most of the necessary data generally would be available in the United States, and domestic firms would have an incentive to provide those data. Even so, if such assistance were provided to all firms in proportion to their emissions before the emission-reduction policy went into effect (rather than just to firms that had substantial emissions), it would be a large and potentially costly undertaking.

Constraints of the WTO Agreements

Direct payments to offset costs resulting from a carbon tax might be considered to be a subsidy. If some or all of the allowances in a domestic cap-and-trade program were given to firms free of charge by the government rather than being sold to them, they also might be considered to be a subsidy, in which case the program would be subject to the provisions of the Agreement on Subsidies and Countervailing Measures (ASCM). One of the legal analyses that CBO reviewed discussed that possibility.[39] That study suggested that whether providing allowances free of charge would be considered a subsidy under the terms of the ASCM was unclear and that no directly relevant jurisprudence existed, but that study concluded that free issuance would probably be ruled to be a subsidy. Another analysis similarly argued that if allowances were given out free of charge and were then sold by the recipients for profit, the ASCM would probably be implicated.[40]

If the transition assistance was ruled to be a subsidy, it would be subject to restrictions contained in the ASCM. Article 3 of the ASCM would then prohibit assistance that was contingent on the amounts of producers' exports or on their use of domestically made rather than foreign-made intermediate inputs. Article 5 in Part III of the agreement would prohibit assistance that was specific to particular firms or industries and caused adverse effects to the interests of other members of the WTO. Adverse effects include injury (such as lost sales) to a domestic industry, nullification or impairment of benefits accruing from the GATT, and serious prejudice.[41] Making the assistance contingent on continued production or employment levels by firms receiving the assistance would create an incentive for increased production and thereby make a finding of adverse effects more likely. Finally, if the assistance was specific to particular firms or industries, Part V of the agreement would allow foreign countries to impose countervailing duties on their imports of the products of those firms if the imports injured or threatened injury to the countries' own industries.

Even so, a country could attempt to defend such subsidies on the grounds that they are intended to address environmental concerns. The ASCM previously contained a provision to allow for certain subsidies based on environmental adaptation. Although that provision has expired, one analyst argued that a country might contend that such an exception is still justified.[42] The resolution of such a claim could take a long time and would ultimately depend on the WTO dispute resolution process. The length of time involved to resolve a challenge relative to the duration of the transition assistance itself would be a consideration in whether a challenge was likely to be made.

End Notes

[1] For additional information, see Congressional Budget Office, *Effects of a Carbon Tax on the Economy and the Environment* (May 2013), www.cbo.gov/publication/44223.

[2] Policies that impose a price on GHG emissions would affect the competitiveness of some producers of services as well as some producers of goods. This report focuses on goods, which are generally more emission-intensive than services. In addition, international trade agreements regarding services differ from those involving goods, so the analysis in this report does not fully apply to services.

[3] For a more detailed discussion of the issues and findings summarized here, see Bruce Arnold, *International Trade and Carbon Leakage*, Working Paper 2013-08 (Congressional Budget Office, December 2013), www.cbo.gov/publication/44970.

[4] Leakage may also occur in connection with reductions in greenhouse gases from activities not subject to limits on emissions. See, for example, Congressional Budget Office, *The Use of Offsets to Reduce Greenhouse Gases* (August 2009), www.cbo.gov/ publication/20972, and *Deforestation and Greenhouse Gases* (January 2012), www.cbo.gov/publication/42686.

[5] For more information on the ability of U.S. fuel consumption to influence the price of carbon fuels in other countries, see Congressional Budget Office, *Energy Security in the United States* (May 2012), www.cbo.gov/publication/43012.

[6] Although the U.S. airline industry has foreign competition on international routes, it would not be put at a competitive disadvantage under a carbon tax or a cap-and-trade program. The reason is that such policies would affect the industry through the price of fuel, and U.S. and foreign airlines would generally purchase their fuel at the same locations and would therefore face the same costs. Hence, no significant carbon leakage would occur in that industry. CBO's calculation is based on estimates of GHG emissions from Environmental Protection Agency, *Inventory of U.S. Greenhouse Gas Emissions and Sinks: 1990–2010*, EPA 430- R-12-001 (April 2012), p. 2-19, Table 2-14, http://go.usa.gov/ T8mj.

[7] Direct emissions are those attributable to the industry's production and to the production of the electricity it uses but not to the production of intermediate inputs the industry purchased. Value added is the value of the industry's output minus the value of the intermediate inputs it purchased.

[8] For more information, see Josh Ederington, Arik Levinson, and Jenny Minier, "Footloose and Pollution-Free," *Review of Economics and Statistics* vol. 87, no. 1 (February 2005), pp. 92–99, www.mitpressjournals.org/toc/rest/87/1.

[9] See Environmental Protection Agency, *EPA Analysis of the Lieberman-Warner Climate Security Act of 2008, S. 2191 in 110th Congress* (March 14, 2008), http://go.usa.gov/T8VR (PDF, 4MB) (S. 2191 was reported from the Senate Committee on Environment and Public Works on May 20, 2008, but no further action was taken on the bill); Liwayway Adkins and others, *The Impact on U.S. Industries of Carbon Prices With Output-Based Rebates Over Multiple Time Frames*, RFF DP 10-47 (Resources for the Future, December 2010), http://tinyurl.com/ c8usx6r; J.P.M. Sijm and others, *Spillovers of Climate Policy: An Assessment of the Incidence of Carbon Leakage and Induced Technological Change Due to CO_2 Abatement Measures*, Report 500036-002 (Netherlands Research Programme on Climate Change, December 2004), pp. 13–14 and 31, http://www.tinyurl.com/lh3f3pw (PDF, 1.29 MB); Jean-Marc Burniaux and Joaquim Oliveira Martins, *Carbon Emission Leakages: A General Equilibrium View*, OECD Economics Department Working Paper 242 (Organisation for Economic Co-operation and Development, May 2000), http://tinyurl.com/ p5tzn4c (PDF, 548 KB); Environmental Protection Agency, *The Effects of H.R. 2454 on International Competitiveness and Emission Leakage in Energy-Intensive Trade-Exposed Industries*, An Interagency Report Responding to a Request From Senators Bayh, Specter, Stabenow, McCaskill, and Brown (December 2, 2009, with corrections on February 23, 2010), http://go.usa.gov/T8pT (PDF, 1.8 MB); and ZhongXiang Zhang, *Competitiveness and Leakage Concerns and Border Carbon Adjustments*, Working Papers in English 2012.080 (Fondazione Eni Enrico Mattei, 2012), http://tinyurl.com/clzqb5z.

[10] Christoph Böhringer, Edward J. Balistreri, and Thomas F. Rutherford, eds., "The Role of Border Carbon Adjustment in Unilateral Climate Policy: Results from EMF 29," *Energy Economics*, vol. 34, Supp. 2 (December 2012), pp. S95–S250, http://tinyurl.com/c65mott.

[11] See ZhongXiang Zhang, *Competitiveness and Leakage Concerns and Border Carbon Adjustments*, Working Papers in English 2012.080 (Fondazione Eni Enrico Mattei, 2012), p. 3, http://tinyurl.com/clzqb5z, which summarizes a number of studies. The estimates in Liwayway Adkins and others, *Carbon Pricing With Output-Based Subsidies: Impact on U.S. Industries Over Multiple Time Frames*, RFF DP 12-27 (Resources for the Future, June 2012), http://tinyurl.com/lnn2d6s, are also consistent with that conclusion.

[12] For more information about the effects of emission-reduction policies on employment, see Congressional Budget Office, *How Policies to Reduce Greenhouse Gas Emissions Could Affect Employment* (May 2010), www.cbo.gov/publication/41257.

[13] Joseph E. Aldy and William A. Pizer, *The Competitiveness Impacts of Climate Change Mitigation Policies*, Working Paper 17705 (National Bureau of Economic Research, December 2011), www.nber.org/papers/w17705.

[14] The estimates cited here are for the effect of H.R. 2454, the American Clean Energy and Security Act of 2009, with an allowance price of $20 per ton but without the provisions of that bill that were designed to offset effects on industry competitiveness. See Environmental Protection Agency, *The Effects of H.R. 2454 on International Competitiveness and Emission Leakage in Energy-Intensive Trade-Exposed Industries*.

[15] Liwayway Adkins and others, *Carbon Pricing With Output-Based Subsidies: Impact on U.S. Industries Over Multiple Time Frames*, RFF DP 12-27 (Resources for the Future, June 2012), http://tinyurl.com/lnn2d6s. The numbers given here are for what the paper estimates in the case of no output-based subsidies.

[16] Besides border adjustments, some papers have examined the possibility of changing the existing tax structure—reducing the corporate income tax rate or lowering the payroll tax rate, for example—to compensate industries whose international competitiveness would be harmed by an emission-reduction program. See, for example, Gilbert E. Metcalf, *Using the*

Tax System to Address Competition Issues with a Carbon Tax, RFF DP 13-30 (Resources for the Future, October 2013), http://tinyurl.com/lrrazve.

[17] However, because border adjustments would reduce leakage and therefore foreign emissions, policymakers might be less concerned if the goal for reducing domestic emissions is missed only by the amount of the increase in domestic emissions resulting from the adjustments.

[18] For further discussion, see Michael O. Moore, *Implementing Carbon Tariffs: A Fool's Errand?* Working Paper IIEP-WP-2010-2 (George Washington University, Institute for International Economic Policy, June 2010), http://tinyurl.com/poyrxgn (PDF, 356 KB).

[19] Those laws are in Title VII of the Trade Act, as added July 26, 1979, P.L. 96-39, as amended; codified at 19 U.S.C. §§1671– 1677n. The most frequently used antidumping law provides for the imposition of duties on imports that are sold at a price below their cost of production or below the price at which the producer sells the good in its home market. The countervailing-duty law provides for duties on imports that have been subsidized by the producer's home government.

[20] Article 6.8 of the Agreement on Implementation of Article VI of the General Agreement on Tariffs and Trade 1994, 1868 U.N.T.S. 201; and Article 12.7 of the Agreement on Subsidies and Countervailing Measures, April 15, 1994, Marrakesh Agreement Establishing the World Trade Organization, Annex 1A, 1867 U.N.T.S. 14.

[21] One of the agreements to which all WTO members are signatories is the General Agreement on Tariffs and Trade 1994, April 15, 1994, Marrakesh Agreement Establishing the World Trade Organization, Annex 1A, 1867 U.N.T.S. 187 (GATT 1994). GATT 1994 consists of the General Agreement on Tariffs and Trade, Oct. 30, 1947, 55 U.N.T.S. 187 (GATT 1947), and various other GATT decisions, waivers, understandings, and provisions as noted in the text of GATT 1994.

[22] Understanding on Rules and Procedures Governing the Settlement of Disputes, April 15, 1994, Marrakesh Agreement Establishing the World Trade Organization, Annex 2, 1869 U.N.T.S. 401.

[23] See Larry Parker and Jeanne J. Grimmett, *Climate Change: EU and Proposed U.S. Approaches to Carbon Leakage and WTO Implications*, CRS Report for Congress R40914 (Congressional Research Service, April 12, 2010); Joost Pauwelyn, *U.S. Federal Climate Policy and Competitiveness Concerns: The Limits and Options of International Trade Law*, Working Paper NI WP 07-02 (Duke University, April 2007), http://tinyurl.com/poyrxgn; Jason E. Bordoff, "International Trade Law and the Economics of Climate Policy: Evaluating the Legality and Effectiveness of Proposals to Address Competitiveness and Leakage Concerns," in Lael Brainard and Isaac Sorkin, eds., *Climate Change, Trade, And Competitiveness: Is a Collision Inevitable?* (Brookings Institution Press, 2009), http://tinyurl.com/cxwfhf4; Gary Clyde Hufbauer, Steve Charnovitz, and Jisun Kim, *Global Warming and the World Trading System* (Peterson Institute for International Economics, March 2009), http://tinyurl.com/bnm3ovd; Robert Howse and Antonia Eliason, "Domestic and International Strategies to Address Climate Change: An Overview of the WTO Legal Issues," in Thomas Cottier, Olga Nartova, and Sadeq Z. Bigdeli, eds., *International Trade Regulation and the Mitigation of Climate Change* (Cambridge University Press, 2009), Chapter 4, http://tinyurl.com/p5upkoj; and Andrew Green, "Climate Change, Regulatory Policy and the WTO: How Constraining Are Trade Rules?" *Journal of International Economic Law,* vol. 8, no. 1 (March 2005), pp. 143–189, http://jiel.oxfordjournals.org/content/8/1/143.

[24] Unless otherwise specified, "Articles" refer to Articles of GATT 1947, as modified, amended, or interpreted upon integration into GATT 1994.

[25] Two analyses discuss the possibility of the allowance requirement for imports being considered as part of an internal regulation applied to all products sold domestically, whether domestically produced or imported, rather than as a tax or import prohibition. Such regulations must be consistent with Article III.4. See Pauwelyn, *U.S. Federal Climate Policy and Competitiveness Concerns,* pp. 23–27; and Parker and Grimmett, *Climate Change,* p. 50.

[26] Pauwelyn, *U.S. Federal Climate Policy and Competitiveness Concerns*, pp. 21–22.

[27] See Hufbauer, Charnovitz, and Kim, *Global Warming and the World Trading System*, pp. 39–46, for a discussion of the issue.

[28] Article II:2(a) stipulates than any domestic tax imposed at the border on imports must be consistent with Article III:2.

[29] However, one analysis presents reasons that border adjustments might be found not to violate national treatment even using that rule for "like products." See Pauwelyn, *U.S. Federal Climate Policy and Competitiveness Concerns,* p. 30.

[30] Various provisions of the GATT contain exceptions to the mostfavored-nation requirement. For example, Article XXIV allows for the creation of free-trade areas among WTO members.

[31] Parker and Grimmett, *Climate Change*, p. 53; Pauwelyn, *U.S. Federal Climate Policy and Competitiveness Concerns,* p. 34; and Green, "Climate Change, Regulatory Policy and the WTO: How Constraining Are Trade Rules?" p. 44.

[32] See, for example, Hufbauer, Charnovitz, and Kim, *Global Warming and the World Trading System*, p. 51, which cites Panel Report, *United States–Standards for Reformulated and Conventional Gasoline*, WT/DS2/R (Jan. 29, 1996), paragraph 6.37.

[33] Pauwelyn, *U.S. Federal Climate Policy and Competitiveness Concerns*, p. 35, which cites Appellate Body Report, United States—Import Prohibition of Certain Shrimp and Shrimp Products, WT/DS58/RW/AB (Oct. 12, 1998), paragraph 133.

[34] Bordoff, "International Trade Law and the Economics of Climate Policy," p. 50.

[35] Ibid., p. 51.

[36] Pauwelyn, *U.S. Federal Climate Policy and Competitiveness Concerns*, p. 37.

[37] All except the fourth issue are discussed (in the same order) in Bordoff, "International Trade Law and the Economics of Climate Policy," pp. 51–54. The fourth and fifth issues are mentioned in Pauwelyn, *U.S. Federal Climate Policy and Competitiveness Concerns*, pp. 38–40. The fifth issue is also mentioned in Green, "Climate Change, Regulatory Policy and the WTO," p. 46.

[38] Parker and Grimmett, *Climate Change*, p. 59.

[39] Hufbauer, Charnovitz, and Kim, *Global Warming and the World Trading System*, pp. 61–63.

[40] Howse and Eliason, "Domestic and International Strategies to Address Climate Change: An Overview of the WTO Legal Issues," p. 56.

[41] "Serious prejudice" is defined in Article 6 of the ASCM. The definition concerns effects on international trade, prices, or world market share that result from subsidization at ad valorem rates higher than 5 percent, subsidies to cover operating losses of an industry, subsidies to cover operating losses of an enterprise other than one-time nonrecurrent measures, or direct forgiveness of debt.

[42] Bordoff, "International Trade Law and the Economics of Climate Policy," pp. 57–58.

In: Unintended Effects of Emissions Reduction ISBN: 978-1-63117-902-0
Editor: Kassy Kinsley

Chapter 2

INTERNATIONAL TRADE AND CARBON LEAKAGE*

Bruce Arnold

ABSTRACT

Under a broad-based carbon tax or cap-and-trade program, some of the reduction in U.S. carbon dioxide emissions would probably be offset by increases in foreign emissions that would not otherwise have occurred, a phenomenon known as carbon leakage. Industries with substantial total emissions, high trade ratios, and high emission intensities are the most likely to generate substantial leakage. Therefore, the industries most likely to be sources of significant leakage through trade in their products are the chemical; primary metal (such as aluminum and iron and steel); and, to a lesser extent, the nonmetallic mineral products (cement, lime, gypsum, and glass) and petroleum and coal products (refining and coke production) industries. Under narrower programs targeting particular industries, significant leakage would occur in fewer industries.

Studies of economywide programs have produced estimates of leakage ranging from 1 percent to 23 percent of the emission reduction the programs would achieve in the countries implementing them. However, those estimates may not apply to future proposals, and estimating leakage is difficult and subject to considerable uncertainty.

---

* This is an edited, reformatted and augmented version of Working Paper 2013-08, released by the Congressional Budget Office, dated December 2013.

INTRODUCTION

Under any policy to reduce carbon dioxide (CO2) emissions that does not provide for subsidies sufficient to offset the cost of the reductions, some industries' or firms' production costs would increase, generally causing those industries or firms to raise the prices of the goods they produce. In the case of broad-based programs that cover all or almost all emissions, the prices of emission-intensive goods (that is, goods whose production requires the emission of large amounts of greenhouse gases relative to the goods' value) would increase the most. For more narrowly targeted programs, the prices of the products of the industries or firms targeted would increase the most. The higher prices would create incentives for other firms and consumers to purchase fewer of those goods, leading to declines in the production of those goods and the resulting emissions.

The higher costs would also make some U.S. producers less competitive relative to producers in countries that do not impose comparable emission-reduction policies. Therefore, some domestically produced goods would be displaced by foreign goods, and that displacement would cause some of the reduction in U.S. emissions to be offset by increases in foreign emissions that would not otherwise have occurred, a phenomenon known as carbon leakage.

Leakage can occur not only through trade but also through another mechanism, namely, through the price of carbon-based fuels. A U.S. emission-reduction program would reduce the use of carbon-based fuels in the United States. Because the fuels are internationally traded (primarily petroleum and, to a lesser extent, coal), the resulting reduction in demand would generally cause their prices in other countries to decline relative to what they would be otherwise. As prices dropped, demand for the fuels in other countries would rise; the amount of that increase would vary by fuel, depending on the ease with which it is traded and other factors. The resulting increase in use of the fuel in those countries would raise their emissions. This paper focuses on where in the economy leakage through trade is likely to be most significant and on the likely magnitude of the total leakage that would occur.

This analysis primarily discusses leakage that might occur under a tax on carbon dioxide emissions or a cap-and-trade program covering such emissions. To the extent that such policies also applied to other greenhouse gases (GHGs), they would be subject to leakage as well. For example, the agricultural sector is a significant source of methane, a potent greenhouse gas, and thus could be a source of leakage under a policy that taxed or applied caps to methane emissions.

Roughly two-thirds of U.S. GHG emissions and three-quarters of U.S. CO2 emissions originate from the residential, commercial, construction, and transportation sectors. Aside from international air travel, those sectors face little international competition. Emission reductions achieved in those sectors would not be subject to being offset by carbon leakage through trade in the products of emission-intensive industries, although they would be subject to leakage through reductions in the international prices of carbon-based fuels (especially petroleum).

The likely levels of leakage in the other sectors—agriculture, mining, and manufacturing—resulting from an economywide tax or cap placed on CO2 emissions vary from industry to industry, depending on several factors. Particularly important are the following three:

- The industry's total CO_2 emissions;
- The amount of foreign competition, which can be roughly measured by the trade ratio (that is, the ratio of exports plus imports to U.S. market size); and
- The percentage increase in the industry's cost imposed by the program, which is likely to be roughly proportional to the industry's CO_2 emission intensity—one measure of which is the ratio of its direct emissions to its value added.

Industries with substantial total CO_2 emissions, high trade ratios, and high CO_2 emission intensities— such as the chemical, mining, and primary metals industries—are the most likely to generate substantial leakage under a broad-based carbon tax or cap-and-trade program that includes no provisions to reduce leakage.

Researchers have produced a number of estimates of the likely amounts of total leakage that would occur under various broad-based emission-reduction policies. Estimates of leakage from the studies that CBO reviewed range from 1 percent to 23 percent of the emission reduction the programs would achieve in the countries implementing them. Recently, a group of 12 modeling teams examined a policy consisting of a 20 percent reduction in emissions from their level in 2004 that was hypothetically implemented by a large group of industrialized countries that included the United States. The leakage rates projected by the modeling teams ranged from 5 percent to 19 percent, with a mean of 12 percent for the estimates. Estimates of leakage from the studies that CBO reviewed are generally for leakage that occurs long after the program begins. Leakage would probably be lower in the short term, and studies of programs in Europe have not found evidence of significant leakage in the first few years. Caution must be exercised in using the estimates to draw conclusions about other emission-reduction proposals, for reasons explained later in the paper.

The studies that produced the long-term estimates generally assumed that the emission-reduction programs included no provisions to mitigate the policies' effects on leakage. Many of them did not distinguish the modes of leakage—that is, through trade or energy prices. Those that did generally found that leakage through energy prices would be more than half of the total leakage, but estimates of leakage through that mechanism are sensitive to the choice of values of some parameters (in particular, the sensitivity of coal production to the price of coal and the extent to which consumers are indifferent about where the coal is obtained) for which estimates in the literature are sparse and uncertain.

As a group, industries whose products are traded internationally produce only a small portion of GHG emissions in the United States. Hence, the loss of market share to foreign competitors and consequent percentage rates of leakage through trade could be much higher for some of those industries than for the economy as a whole. How those rates would compare with the estimated rates that are presented here for the economy, which include leakage through trade and leakage through the prices of carbon-based fuels, would depend on how much of the leakage occurs through fuel prices.

DETERMINING WHERE SUBSTANTIAL LEAKAGE MIGHT OCCUR

For an economic sector or industry to be the source of substantial carbon leakage through trade in its products, it must:

- Produce sizable quantities of carbon dioxide;
- Have significant foreign competition not subject to the U.S. emission-reduction program or to an equivalent program implemented in its home country at the same time as the U.S. program; and
- Face a substantial percentage increase in its costs as a result of the emission-reduction program.

The following analysis pertains to an economywide carbon tax or cap-and-trade program. The analysis and conclusions assume that the emission-reduction program has no border adjustments or other provisions (such as transition assistance linked to changes in output or employment) that are designed to, or that might be expected to, prevent or ameliorate the reduction in competitiveness that causes the leakage.

The same industries identified by the analysis as likely sources of leakage would also be likely sources under a more narrowly focused carbon tax or cap-and-trade program to the extent that the program targeted those industries. If a program targeted only a subset of those industries, then only that subset would be a likely source of leakage.

Similarly, the same industries would be likely sources of leakage under standards regulating production processes to the extent that those standards targeted and raised the costs of those industries. If the standards targeted only a subset of the industries, then only that subset would be likely sources of leakage.

If the standards targeted other industries, less leakage would be likely to occur because either the amount of foreign competition or the emissions of those industries (or both) would be lower.

Significant leakage through trade would not be likely for standards regulating products, such as fuel economy standards for automobiles. The reason is that imported products would have to meet the same standards as domestically produced products, and exported domestic products would not have to meet the standards. Thus, the standards would not significantly affect the competitiveness of domestic production relative to foreign production and therefore would have little, if any, effect on trade. The analysis and conclusions in this paper do not apply to such standards.

The analysis and conclusions also do not apply to policies that include or take the form of subsidies. An example of such a subsidy is the tax credits under current law for the use of renewable sources of energy by producers of electricity.[1] Subsidies would offset the cost to firms of emission reduction and could result in lower rather than higher prices. Hence, their effects on trade could be different from the effects of other emission-reduction policies.

Broad Economic Sectors Not Subject to Trade-Related Leakage

To analyze the effects of emission-reduction programs, the most useful approach to quantifying the emissions of various parts of the economy is to attribute the emissions of the electrical generating sector (which accounted for 34 percent of all GHG emissions in the United States in 2010) to the other sectors or industries in proportion to their use of electricity and then adding those attributed emissions to the sectors' or industries' own emissions from production. The resulting sum for each sector or industry is commonly called its *direct emissions.*

The direct emissions of the transportation, residential, and commercial sectors accounted for about 62 percent of all GHG emissions in the United States in 2010—27 percent, 18 percent, and 17 percent, respectively (see Table 1). Those sectors face little foreign competition that would not be covered by a U.S. emission-reduction program.[2] They would therefore not be a source of leakage through trade in their products. The same is true for the construction sector, but a separate estimate of that sector's GHG emissions was not available.

Table 1. Direct Emissions of Greenhouse Gases, by U.S. Economic Sector, 2010

(Percent)			
Sector	All Greenhouse Gases		Carbon Dioxide Only
Agriculture	8	1	
Mining, Construction, and Manufacturing	30	29	
Transportation	27	31	
Residential	18	21	
Commercial	17	18	
Total	100	100	

Source: Congressional Budget Office based on estimates of greenhouse gas and carbon dioxide emissions from Environmental Protection Agency, *Inventory of U.S. Greenhouse Gas Emissions and Sinks: 1990– 2010*, EPA 430-R-12-001 (April 2012), p. 2-19, Table 2-14, http://go.usa.gov/T8mj.

Notes: A sector's direct emissions are the sum of its emissions and the emissions attributable to the production of the electricity it uses.

The data source includes separate estimates for the emissions from U.S. territories. CBO excluded those estimates when calculating the numbers in this table.

The same four sectors—transportation, residential, commercial, and construction—produced 72 percent of CO_2 emissions in the United States. The agricultural, mining, and manufacturing sectors, which account for 28 percent of CO_2 emissions, produce goods that are traded internationally and therefore could become sources of carbon leakage through trade (that total excludes emissions from the construction industry from the total reported in Table 1 for emissions from mining, construction, and manufacturing).[3] The rest of this section discusses which industries in those sectors would probably generate the most carbon leakage through trade.

Analysis of the Agriculture, Mining, and Manufacturing Sectors

Within the sectors that face international competition, the likelihood of sizable leakage varies substantially among industries because of differences in the extent of their emissions, the foreign competition they face, and the amounts by which an emission-reduction program would raise their costs.

The industries most likely to experience large effects under an emission-reduction program can be identified by considering each of those characteristics in turn.

Industries with Large Emissions

Of the 23 industries in the agricultural, mining, and manufacturing sectors, each of the top nine emitters (ranked by direct emissions) was the source of between 1 percent and 6 percent of direct emissions of CO_2 in the United States in 2010. Those industries are petroleum and coal products (refining); chemicals; mining; primary metals; nonmetallic mineral products; food; agriculture, forestry, and fisheries; paper; and computer and electronic products (see Table 2). Each of the bottom nine emitters produced only 0.3 percent or less. Hence, all else being the same, the top nine emitters would be much more likely to be the source of sizable leakage through trade in their products than would the bottom nine.

Examining industries' *total emissions* of CO_2—that is, direct emissions plus the emissions entailed in producing the intermediate inputs the industries purchase for producing their goods—does not change the picture much. The nine industries with the largest emissions are mostly the same; the rank order shifts slightly, and the paper industry is replaced by the transportation equipment industry. Each industry was the source of between 1.8 percent and 8.5 percent of total emissions of CO_2 in the United States.

The analysis presented here focuses on CO_2 emissions because of a lack of data for other GHG emissions at the industry level. Carbon dioxide makes up more than four-fifths of GHG emissions in the United States, however, so the numbers for CO_2 emissions should be similar to those for GHG emissions for most industries.

Table 2. CO_2 Emissions in the U.S. Agricultural, Mining, and Manufacturing Sectors, by Industry, 2010

(Percentage of all CO_2 emissions)		
Industry	Direct Emissions[a]	Total Emissions[b]
Petroleum and Coal Products	6.2	8.5
Chemicals	4.5	5.9
Mining	3.4	3.7
Primary Metals	2.7	3.1
Nonmetallic Mineral Products	2.1	2.3
Food	1.7	3.8
Agriculture, Forestry, and Fisheries	1.5	2.1
Paper	1.1	1.4
Computer and Electronic Products	1.0	2.1
Plastic and Rubber Products	0.8	1.5
Fabricated Metal Products	0.7	1.4
Transportation Equipment	0.6	1.8
Machinery	0.6	1.7
Wood Products	0.4	0.7
Textile Mills	0.3	0.3
Electrical Equipment, Appliances, and Components	0.2	0.5
Printing and Related Support	0.2	0.5
Miscellaneous Manufacturing	0.2	0.6
Beverage and Tobacco Products	0.2	0.7
Furniture and Related Products	0.1	0.3
Textile Mill Products	0.1	0.2
Apparel	*	0.1
Leather and Allied Products	*	*

Source: Congressional Budget Office estimates based on data for industry emissions in 2006 from Department of Commerce, Economics and Statistics Administration, *U.S. Carbon Dioxide Emissions and Intensities Over Time: A Detailed Accounting of Industries, Governments and Households* (April 2010), http://go.usa.gov/T8P9; data for sector emissions in 2006 and 2010 from Environmental Protection Agency, *Inventory of U.S. Greenhouse Gas Emissions and Sinks: 1990–2010*, Report No. 430-R-12-001 (April 2012), p. 2-19, Table 2-14, http://go.usa.gov/T8mj; and industry value-added, output, and price data from the Bureau of Economic Analysis.

Note: CO_2 = carbon dioxide; * = between zero and 0.05 percent.

[a] An industry's direct emissions are the sum of its emissions and the emissions attributable to the production of the electricity it uses.

[b] An industry's total emissions are the sum of its direct emissions and the emissions attributable to the production of the intermediate inputs it uses.

A notable exception is the agriculture, forestry, and fisheries industry, whose direct emissions accounted for only 1.5 percent of all CO_2 emissions in the United States in 2010 but for 7.7 percent of all GHG emissions in that year.

The discrepancy arises because 84 percent of the industry's GHG emissions were gases other than CO_2 (mostly methane and nitrous oxide).[4] The cap-and-trade proposals the Congress has considered have not capped any of the non-CO_2 greenhouse gas emissions of that industry. Hence, the likelihood of sizable carbon leakage in that industry under the proposed programs would be a function of the sector's emissions of CO_2, which are not particularly high, rather than a function of its higher emissions of all greenhouse gases. However, if the United States were to tax or place a cap on emissions of methane and nitrous oxide in addition to CO_2, significant leakage could occur in the agricultural sector.

Industries That Have Substantial International Competition

The extent of foreign competition an industry faces can be roughly measured by the dollar values of its imports and exports relative to the size of the U.S. market. For a number of reasons, those ratios are imperfect indicators of the degree of competition. For example, some industries ship parts produced in the United States to low-wage countries such as Mexico or China for labor-intensive final assembly and then import the assembled products back to the United States.

In such cases, the exports, imports, and domestic production may all be classified as belonging to the industry in question, leading to high trade values for the industry; yet the factories supplying the imports are not competing with U.S. producers. Nevertheless, a comparatively high ratio of imports or of exports to the size of the U.S. market indicates that a good is more easily traded than are other goods, thereby raising the possibility that an emission-reduction program might lead to a loss in market share and subsequent carbon leakage for the industry producing the good.

Comparatively low ratios of imports and of exports of a good suggest that it is less easily or economically traded than are other goods and that little leakage is therefore likely to result from trade in that good.

As measured by U.S. trade with all countries, a large fraction of the 23 industries in the agricultural, mining, and manufacturing sectors face sizable foreign competition. Eleven of the industries—leather and allied products; apparel; machinery; computer and electronic products; electrical equipment, appliances, and components; miscellaneous manufacturing; transportation equipment; textile mills; chemicals; primary metals; and textile mill products—have *total trade* (exports plus imports) exceeding 50 percent of their respective U.S. market sizes (see Table 3).

Table 3. Ratio of U.S. Trade to U.S. Market Size in the Agricultural, Mining, and Manufacturing Sectors, by Industry, 2010

(Percent)						
	Trade with All Countries			**Trade with Non-Annex II Countries**[a]		
Industry	**Exports**	**Imports**	**Total**	**Exports**	**Imports**	**Total**
Leather and Allied Products	7	96	103	5	93	97
Apparel	3	87	91	2	86	87
Machinery	43	38	80	27	16	43
Computer and Electronic Products	20	57	77	13	48	62
Electrical Equipment, Appliances, and Components	21	46	67	12	36	48
Miscellaneous Manufacturing	18	47	65	7	35	43
Transportation Equipment	27	37	64	14	15	28
Textile Mills	30	26	56	24	17	41
Chemicals	27	28	55	13	10	23
Primary Metals	20	32	52	8	17	25
Textile Mill Products	7	45	52	3	42	45
Mining	4	38	42	2	29	31
Furniture and Related Products	4	33	38	1	29	31
Plastic and Rubber Products	13	19	32	6	12	18
Agriculture, Forestry, and Fisheries	19	13	32	13	10	23
Fabricated Metal Products	11	17	28	6	10	16
Nonmetallic Mineral Products	9	18	27	4	12	16
Paper	13	13	27	7	4	12
Petroleum and Coal Products	9	15	24	7	10	17
Wood Products	7	17	23	3	8	11
Beverage and Tobacco Products	4	12	15	2	4	5
Food	8	6	14	5	3	8
Printing and Related Support	7	6	13	2	4	6

Source: Congressional Budget Office based on trade data from the Bureau of the Census and input-output data from the Bureau of Economic Analysis.

Note: The U.S. market for each industry is defined as U.S. industry output plus imports minus exports. Export numbers are free-alongside-ship values (the value of a commodity at the port of exportation, generally including the purchase price plus all charges incurred in placing the commodity alongside the carrier at the port of exportation in the country of exportation). Import numbers are cost, insurance, and freight (sometimes referred to as CIF) values. Industry output data are from the use tables of the input-output accounts published by the Bureau of Economic Analysis.

[a] Countries that signed Annex II of the U.N. Framework Convention on Climate Change thereby agreed to an additional provision of the convention requiring those countries to provide financial resources to cover the costs incurred by developing countries in meeting their obligations under the convention. Non-Annex II countries—that is, countries that did not sign Annex II—might be considered less likely than Annex II countries to implement emission-reduction programs comparable with one that the United States might implement.

One might argue that an industry's trade with all countries is not the best measure to use for assessing the likelihood of significant leakage. If all or most of the industry's foreign competitors were in countries that imposed comparable emission-reduction programs of their own, those competitors would face increases in cost similar to those faced by firms in the U.S. industry. Consequently, the competitiveness of the U.S. industry would change little if at all. Therefore, the argument would go, a better indicator of competition might be trade with countries that are unlikely to impose comparable emission-reduction programs of their own.

Although one can make counterarguments, the proposed alternative measure is nevertheless informative. It is not possible to predict with confidence which U.S. trading partners would adopt comparable policies, but a country's not having signed on to Annex II of the U.N. Framework Convention on Climate Change may be taken as an indicator that the country is less likely to implement comparable policies than is a signatory country.[5] That list, however, is only a rough indicator. Some countries that signed Annex II ultimately may not undertake action, and some that did not sign Annex II may undertake action.

A further qualification to the use of a country's signing or not signing onto Annex II as such an indicator is that some signatories—notably the European Union (EU), Australia, and New Zealand—have already begun (or have decided to begin) implementing emission-reduction programs. Therefore, if the United States implemented a program, those countries would not be doing so at the same time as the United States. In the absence of a U.S. program, the programs in those countries could be expected to cause (and may in some cases already be causing) changes in U.S. trade with those countries that would create leakage for those programs. That is, U.S. exports of the products of emission-intensive industries to those countries would increase and imports of such products would decline—the opposite of what would result from a U.S. emission-reduction program. If the United States then implemented such a program, those changes in trade would be reversed. The competitiveness of U.S. firms in emission-intensive industries would decline relative to the competitiveness of firms in those countries, and leakage would occur, compared with the status quo in which the other countries were implementing their programs but the United States was not.

Competitiveness would not decline, however, nor would any leakage occur compared with what would be the case if neither the United States nor the other countries implemented programs.

That qualification being noted, roughly the same industries stand out as having more international competition than the others regardless of whether the competition is measured by trade with all countries or by trade with non-Annex II countries (see Table 3). The two industries with the highest trade ratios— leather and allied products and apparel—have values of trade with non-Annex II countries that are not much lower than the values with the world as a whole, which indicates that almost all of the foreign competition for the industries is in countries less likely than others to implement their own emission-reduction programs. For the rest of the industries in the top 11 (those whose ratios of trade with the world to U.S. market size are at least 50 percent), the differences between trade with all countries and trade with non-Annex II countries are larger—much larger in a number of cases—although trade with non-Annex II countries is still sizable enough to allow some leakage to occur.

Industries with both sizable CO2 emissions and large ratios of trade to U.S. market size include the chemicals, mining, primary metals, and computer and electronic products industries. (Mining ranks 12th in the ratio for trade with the world, but 9th—above chemicals and primary metals—for trade with non-Annex II countries.) Compared with those four industries, the machinery and transportation equipment industries have trade ratios that are similarly large but emissions that are somewhat lower; the agriculture, forestry, and fisheries industry has similarly large emissions but somewhat smaller trade ratios; and the petroleum and coal products (refining and coke production) industry and the nonmetallic mineral products industry have sizable emissions but even smaller trade ratios.

Product and industry characteristics raise the likelihood of significant leakage in the chemicals and primary metals industries relative to that in the transportation equipment and computer and electronic products industries. The chemicals and primary metals industries tend to produce fairly homogeneous products, giving purchasers little reason to care whether they buy U.S. products or competing products made by foreign producers. That fact would suggest a higher likelihood of significant leakage. By contrast, the products of the transportation equipment industry and possibly the computer and electronic products industry are more heterogeneous, which may reduce the likelihood of significant leakage. Those two industries also in some cases assemble products in some countries from parts produced in other countries, which may inflate the trade ratios.

Industries Whose Costs Would Rise Substantially

Simply having sizable emissions and significant exposure to foreign competition would not necessarily cause an industry to become less competitive as a result of an emission-reduction program. An industry with significant foreign competition would suffer competitively in proportion to the percentage by which its costs increased as a result of the program. Under a program that imposed costs on all or almost all CO_2 emissions—such as an economywide carbon tax or cap-and-trade program—industries' costs would increase by an amount that is proportional to CO_2 emissions, so a large quantity of emissions would mean a large increase in costs.

Note, however, that the quantity of an industry's emissions might be large simply because the industry itself is large, has a sizable output, and therefore has high total costs. In that case, the percentage increase in cost would be large only if the industry's CO_2 emissions were high relative to the value of its output— that is, if it had a high *emission intensity*.

Two measures of emission intensity are relevant:

- The *direct emission intensity* is the ratio of the sum of direct emissions from fossil-fuel combustion, process emissions, and emissions from electricity consumed by the industry to the value added by the industry.[6]
- The *total emission intensity* is the ratio of the sum of the components of direct emission intensity and emissions from the production of intermediate inputs used by the industry to the value of the industry's gross output.

Which measure is more relevant for a given industry depends on how widespread the emission-reduction program is (for example, whether it affects only the industry or both the industry and its suppliers). It also depends on whether the intermediate inputs are easily traded, and, if they are, whether significant suppliers of the inputs are located in countries not imposing their own emission-reduction programs (or are likely to be located in such countries in the future). However, the industries in the agricultural, mining, and manufacturing sectors with the highest CO_2 emission intensities are the same by either measure, although ordered a bit differently: primary metals, nonmetallic mineral products, petroleum and coal products, textile mills, paper, chemicals, wood products, and mining (see Table 4).

The industries most likely to incur significant leakage are those with sizable emissions, sizable foreign competition, and high emission intensities. The chemicals and primary metals industries have those characteristics (see Figure 1). Petroleum and coal products and nonmetallic mineral products industries produce large quantities of CO_2 emissions and have high emission intensities, but they have less competition as measured by trade.

ESTIMATES OF LEAKAGE

A number of studies have estimated carbon leakage for various economywide emission-reduction programs. Those studies can be grouped into three categories:

- Recent studies that used economic models to project leakage in the long term for specific proposals for reducing U.S. emissions;
- Earlier studies of the Kyoto Protocol that used economic models to project leakage for that agreement; and
- Several studies that used other methodologies to determine the leakage that occurred in the first few years after emission-reduction programs were implemented in Europe and one study that examined the effects on trade of changes in energy prices in the United States that were equivalent in effect to imposing a carbon tax.

Table 4. CO_2 Emission Intensities in the U.S. Agricultural, Mining, and Manufacturing Sectors, by Industry, 2010

Industry	Direct Intensity[a]	Total Intensity[b]
Primary Metals	3.6	0.8
Nonmetallic Mineral Products	3.3	1.5
Petroleum and Coal Products	2.0	0.7
Textile Mills	1.8	0.7
Paper	1.1	0.5
Chemicals	1.1	0.5
Wood Products	1.0	0.6
Mining	0.8	0.5
Plastic and Rubber Products	0.6	0.5
Food	0.6	0.3

Industry	Direct Intensity[a]	Total Intensity[b]
Agriculture, Forestry, and Fisheries	0.5	0.3
Textile Mill Products	0.4	0.4
Printing and Related Support	0.3	0.3
Fabricated Metal Products	0.3	0.3
Transportation Equipment	0.3	0.2
Electrical Equipment, Appliances, and Components	0.3	0.2
Machinery	0.2	0.3
Computer and Electronic Products	0.2	0.3
Furniture and Related Products	0.2	0.3
Beverage and Tobacco Products	0.2	0.3
Apparel	0.2	0.3
Leather and Allied Products	0.2	0.4
Miscellaneous Manufacturing	0.1	0.2

Source: Congressional Budget Office estimates based on data for industry emissions in 2006 from Department of Commerce, Economics and Statistics Administration, *U.S. Carbon Dioxide Emissions and Intensities Over Time: A Detailed Accounting of Industries, Governments and Households* (April 2010), http://go.usa.gov/T8P9; data for sector emissions in 2006 and 2010 from Environmental Protection Agency, *Inventory of U.S. Greenhouse Gas Emissions and Sinks: 1990–2010*, EPA 430-R-12-001 (April 2012), p. 2-19, Table 2-14, http://go.usa.gov/T8mj; and industry value-added, output, and price data from the Bureau of Economic Analysis.

Note: CO2 = carbon dioxide.

[a] An industry's direct intensity is the ratio of its direct emissions to its value added—that is, to the value of its output minus the value of the intermediate inputs it uses to produce the output. (Direct emissions are the sum of the industry's emissions and the emissions attributable to the production of the electricity it uses.) Direct intensity is measured here in metric tons of CO2 per thousand dollars of value added.

[b] An industry's total intensity is the ratio of its total emissions to the value of its output. (Total emissions are the sum of an industry's direct emissions and the emissions attributable to the production of the intermediary inputs it uses.) Total intensity is measured here in metric tons of CO2 per thousand dollars of total output.

Caution must be exercised in using the estimates from such studies to draw conclusions about other proposals, for several reasons. First, estimates of leakage are subject to considerable uncertainty, in part because analysts must make assumptions about important factors about which there is insufficient knowledge. Second, the reduction in competitiveness of U.S. producers and the consequent amount of leakage that would result from an emission-reduction policy would depend on the actions of other countries.

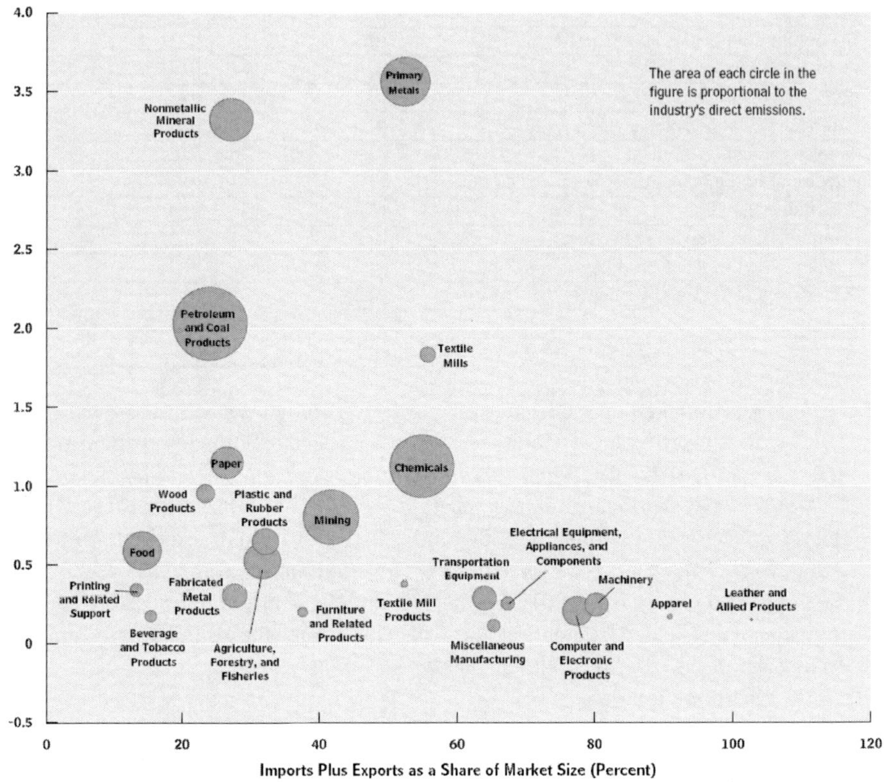

Source: Congressional Budget Office.

Notes: The largest amounts of leakage through trade are likely to occur in industries that have sizable emissions of carbon dioxide (CO_2), have high direct emission intensities, and face significant foreign competition as measured by the ratio of an industry's trade—imports plus exports—to the size of the U.S. market. The size of the U.S. market for each industry is measured here by the industry's output plus imports minus exports. As such, the ratio can exceed 100 percent.

Direct CO_2 emissions are the sum of an industry's own emissions and the emissions attributable to the production of the electricity it uses.

The CO_2 emission intensity shown is the direct intensity, which is the ratio of an industry's direct emissions to its value added (that is, to the value of its output minus the value of the intermediate inputs it uses), measured here in metric tons of CO_2 per thousand dollars of value added.

Figure 1. Carbon Dioxide Emission Intensities, International Trade, and Emissions, by Industry, 2010. (Carbon dioxide emission intensity).

For example, if all major U.S. trading partners implemented their own equivalent policies, then an emission-reduction program would not substantially reduce the competitiveness of U.S. producers, and the carbon leakage that would occur through international trade would be relatively small; to the extent that major trading partners did not implement their own policies, the decline in competitiveness and the resulting carbon leakage would be larger. Third, leakage would probably be low in the first few years after a program started and would grow over time. Fourth, the estimates above are for economywide emission-reduction programs such as a carbon tax or a cap-and-trade program. Narrower programs—such as a cap-and-trade program for electric utilities only, production efficiency standards for selected industries, or standards setting minimum requirements for use of renewable fuels by some industries— could have higher or lower rates of leakage, depending on whether the industries or economic sectors they targeted were among those likely to constitute significant sources of leakage.

Finally, the estimates generally do not account for policies that are designed to reduce leakage or that would reduce leakage as one of their effects. Examples of such policies are border adjustments, such as tariffs or subsidies, or transition assistance that is linked to continuing levels of output or employment by the firm receiving it.

Forward-Looking Estimates for U.S. Emission-Reduction Proposals

Several studies have used general-equilibrium modeling to project the amount of leakage that would occur in the long term under various proposals for reducing U.S. emissions.

An analysis by the Environmental Protection Agency (EPA) examined the potential for reduced competitiveness and carbon leakage resulting from the cap-and-trade program for U.S. emissions contained in the Climate Security Act of 2008 (S. 2191).[7] In one scenario, EPA assumed that U.S. efforts to lower GHG emissions would be mirrored by all signatories to the Kyoto Protocol except Russia but that all other countries would take no action. The analysis estimated the likely carbon leakage at values near 10 percent in 2030 and 2050.

Another study estimated leakage under a tax of $15 per ton of CO_2 imposed by all Annex I countries.[8] (Annex I countries are the 38 countries that agreed to limit their emissions in accordance with the U.N. Framework

Convention on Climate Change.) The study projected a long-run leakage rate of 9 percent average over all Annex I countries. It projected the largest leakage rates among manufacturing industries to be for the petroleum and coal products industry (27 percent); the chemicals, rubber, and plastics industries (11 percent); the nonferrous primary metals industry (9 percent); the ferrous metals industry (8 percent); and the nonmetallic mineral products industry (6 percent). Those rates reflect the percentage of the reduction in emissions in each industry that would be offset by increased emissions by competitors in foreign countries.

The same study also addressed the question of how much of total leakage would occur through trade in emission-intensive goods compared with how much how much would occur as a result of the reduction in international prices of carbon-based fuels. That question is important because leakage by the latter mechanism cannot be eliminated or even reduced by measures regulating international trade (aside from a prohibition on the export of coal). A survey of studies estimating leakage for the Kyoto accords (discussed below) noted that there is little empirical evidence on the issue. It noted further that most modelers seem to agree that the energy-price channel is the most important in the short to medium run but that trade in emission-intensive goods might become dominant in the long run as a result of changes in the destination of international investment and the relocation of industries.[9]

Another survey of studies (including studies of European and possible U.S. programs) stated, "In virtually all applied economic models incorporating both channels, the larger part of carbon leakage occurs through this fossil fuel channel, rather than via the competitiveness channel ..."[10] The survey cited five studies to support that statement. Estimates from the study of a tax of $15 per ton of CO_2 are consistent with the statement. The study estimated that 6 percent of the increase in emissions of other countries would be attributable to greater industry output in those countries (leakage through international trade) and that the rest would be attributable either to increased emissions by households or to increased emissions of industry per unit of output (leakage through the price of carbon-based fuels).

Estimates of leakage through the price of carbon-based fuels are sensitive to the values assumed for various parameters in the economic model used to produce the estimates, and the correct values of some of those parameters are not known with much accuracy. One study used a simplified version of a model in the literature to examine the sensitivity of leakage estimates to the assumed values of various parameters.[11] Burniaux and Martins found that estimates of leakage through energy prices were particularly sensitive to the assumed sensitivity of coal production to the price of coal and to the assumed extent to which purchasers of coal were indifferent to its origin—that is, their assumed willingness to switch to coal from another country in response to relative price changes. Thus, if the amount of coal produced was not sensitive to price, a decline in the demand for coal in, for example, the United States as a result of an emission-reduction program would lead to a domestic surplus of coal and a consequent decline in its price. If purchasers in other countries were relatively indifferent about the origins of coal (whether from the United States or some other location), the now-lower U.S. price would prompt them to purchase the surplus U.S. coal, resulting in sizable leakage. However, if the amount produced was sensitive to the price, then the decline in demand in the United States would lower the price of coal, and, in response, the amount of coal produced would decline, resulting in little leakage.

Burniaux and Martins also reviewed the literature for estimates of the sensitivity of production to price and found that such estimates were sparse and mixed.[12] They concluded that leakage is likely to be small for the range of parameters most frequently quoted in the literature but argued that further empirical work was needed on that issue.

Still another study CBO examined estimated the likely leakage from the emission-reduction program in the proposed American Clean Energy and Security Act of 2009 (H.R. 2454).[13] That bill contained provisions to offset the effects of the program on the competitiveness of industries.

The study found that without those provisions the leakage through trade in the products of "trade vulnerable" industries would be on the order of 10 million metric tons of CO_2 equivalent, or about 10 percent of the estimated emission reductions achieved in those industries, compared with leakage of about 1 million metric tons (or about 1 percent) with those provisions. Those numbers cannot be compared with the other numbers presented here for at least three reasons.

First, they are only for a subset of all industries, not for the economy as a whole. Second, they include only leakage through trade and not leakage through the prices of carbon-based fuels. Third, the estimates are for what the study calls the medium term, whereas the other estimates presented here are for the long term. The study does not present any estimates for leakage through both mechanisms for the entire economy.

Forward-Looking Estimates of Leakage under Multinational Emission-Reduction Proposals

Recently, a group of 12 modeling teams projected the leakage that would occur in one particular emission-reduction policy.[14] The policy consisted of a 20 percent reduction in emissions from their level in 2004 that the modelers assumed would be implemented by industrialized countries as listed in Annex I of the U.N. Framework Convention on Climate Change, including the United States but not the Russian Federation. The leakage rates projected by the 12 teams ranged from 5 percent to 19 percent, with a mean of 12 percent for the 12 estimates.

The estimates are sensitive to the size of the coalition of countries assumed to implement the emission-reduction policy. When China was added to the coalition, the mean leakage estimate was only 7 percent rather than 12 percent. When the coalition was reduced to consist only of the countries of the EU and the European Free Trade Area, the mean leakage estimate was 24 percent.

The estimates are also sensitive to what is assumed about the behavior of oil-producing countries. When those producers were assumed to counteract the decline in fuel prices by restricting their output, the mean leakage estimate dropped from 12 percent to less than 3 percent. The reason is that the portion of the 12 percent that represented leakage through the price of petroleum was reduced or eliminated.

In earlier studies, a number of analysts considered the effects of the Kyoto Protocol. That protocol was an international agreement that set emission-reduction targets for many industrialized countries at 5 percent below the 1990 levels, to be achieved between 2008 and 2012, but it did not specify how the targets were to be reached. The United States signed the protocol in 1998, agreeing to reduce GHG emissions by 7 percent from 1990 levels. However, the President never submitted the protocol to the Senate for ratification, and the United States has not attempted to meet its emission-reduction target. Nevertheless, a number of the analyses were conducted under the assumption that the United States would participate.[15]

According to a survey of studies that estimated leakage for Kyoto climate policies under that assumption, most estimates were between 5 percent and 20 percent of the projected emission reductions in Annex I countries.[16] Another study cited estimates in the literature ranging from 2 percent to 21 percent.[17] Still another survey cites other surveys finding estimates ranging from 2 percent to 23 percent.[18] That survey also points to a World Bank study that finds leakage of only 1 percent when high-income countries unilaterally cut their emissions in 2020 by 17 percent relative to their levels in 2005.[19]

Backward-Looking Estimates of Leakage

Other studies have looked at programs after they were implemented to determine how much leakage actually occurred, and another study analogously looked back at the effects of changes in energy prices (equivalent to increases in energy costs imposed by a carbon tax) on U.S. production and consumption and the implications for changes in trade and leakage.

Estimates of Leakage Shortly after Implementation

Even after a program has gone into effect, leakage cannot be observed directly; as is the case for projections of future leakage, current or past leakage must be estimated using models or statistical analysis. Projections of future leakage involve comparing trade at the time the projection is made with what trade is projected to be under the emission-reduction program. Estimates produced after the fact involve comparing trade under the program with what models or statistical analysis project that trade would have been without the program. Either way, known levels of trade must be compared with unknown levels that are projected using modeling or statistical analysis.

The EU implemented a relatively broad-based emission-reduction program known as the Emissions Trading Scheme, or EU-ETS, in 2005. The EU-ETS is a cap-and-trade program that in its pilot period (2005 through 2007) covered about half of the EU's CO2 emissions and 40 percent of its GHG emissions. One study examined several aspects of the program during that period, including the rates of leakage for four industries—cement, iron and steel, oil refining, and aluminum.[20] (The study presented its own estimates and estimates from other studies.) Using statistical analysis, the study estimated net imports of the products as a function of domestic consumption and the price of emission allowances and found the allowance price not to be a significant determinant. Although projections and other evidence beforehand indicated that all four industries were likely to experience significant leakage, the study found no evidence of leakage from 2005 through 2007.

Another study examined the effects of the Climate Change Levy in the United Kingdom, which was begun in 2001.[21] That program imposes taxes on energy ranging from 6.1 percent on coal to 16.5 percent on natural gas. Some businesses are offered an 80 percent reduction in the tax if they join a climate change agreement, under which they must meet targets for energy use or CO2 emissions. The study statistically analyzed differences in terms of output and other variables between firms that joined such agreements and those that did not. Using data through 2004, the study found no significant evidence of leakage.

Still another study examined the responsiveness of U.S. production and consumption of manufactured goods in the short term to changes in energy prices in recent years.[22] That approach is analogous to looking back on the effects of a broad-based emission-reduction program because such programs work by effectively making the use of carbon-based energy more expensive. The study used statistical analysis to determine the extent to which production and consumption of the products of more than 400 manufacturing industries had been affected in recent years by changes in energy prices. It estimated that a change in energy prices that was the equivalent of imposing a tax of $15 per ton on CO2 caused average declines of 1.4 percent in production and 0.4 percent in consumption over the first few years. Those numbers imply that 1 percent of production would shift to other countries in the first few years, thereby producing leakage. The magnitude of the leakage cannot be determined from the results of the study. Nevertheless, unlike the studies of the EU-ETS and the Climate Change Levy, this study found evidence of increases in energy costs (such as would result from a carbon tax) causing displacement of U.S. production by foreign production in the short term.

Estimates of Leakage over the Short and Long Term

The results of the study of the effects of higher energy prices on production and consumption indicate that leakage would probably occur in the early years of an emission-reduction program, and the results of the other two backward-looking studies constitute evidence that leakage at that point would probably be small, perhaps so small as to be difficult to detect. The latter results do not, however, indicate the likely magnitude of leakage in later years.

As noted in the cautions about using leakage estimates, leakage in the initial years after a program is implemented is likely to be much lower than in later years. Even if and when the increase in costs resulting from an emission-reduction program is sufficient to make production less expensive in another country for a given firm, that firm is not likely to move its production until it is convinced the policy will remain in place—that is, that policymakers will neither eliminate the program nor decrease the magnitude of the required reduction in emissions when they face complaints from firms and their employees who lose jobs. In addition, a firm would want to be assured that a similar emission-reduction program would not be implemented in the country to which it intended to move production. That firm also is unlikely to move production until its plant and equipment have largely depreciated. Even if competition from the other country drove down the domestic price to a point that the firm was losing money, that firm would continue to produce as long as its plant and equipment was still functional and the price was higher than the firm's marginal cost (its total cost minus the cost of the plant and equipment, which is a sunk, or unrecoverable, cost). Moreover, certain features of the way that emission allowances were distributed in the pilot period of the EU-ETS could be expected to further delay the movement of production abroad. The resulting low levels of leakage in the early years of a program could be difficult to detect with only a few years of data, but the larger leakage to be expected later would be more easily detected in many years of data.

Estimates of Leakage Rates under More Narrowly Focused Programs

The leakage estimates presented above are for economywide emission-reduction programs such as a carbon tax or a cap-and-trade program applied to all or almost all emissions. Leakage could be different in programs that are more narrowly focused. For example, rates of leakage are likely to be lower for production standards targeting industries with little foreign competition than for such standards targeting industries with substantial foreign competition. Similarly, under a cap-and-trade program that applied only to electrical generating plants, the array of industries facing significant increases in their costs and consequent declines in their competitiveness would be different from that for an economywide cap-and-trade program. Given the differences among industries in the amount of foreign competition they face, the rate of leakage for a cap-and-trade program applied to electrical generating plants would therefore probably be different from that of an economywide program.

End Notes

[1] For additional information, see the testimony of Terry M. Dinan, Senior Advisor, Congressional Budget Office, before the Subcommittee on Energy of the House Committee on Science,

Space, and Technology, *Federal Financial Support for Fuels and Energy Technologies* (March 13, 2013), www.cbo.gov/publication/43993.

[2] Domestic airlines face foreign competition on international flights, but foreign airlines departing from U.S. airports would have to purchase fuel in the United States and would therefore incur the cost of higher prices for such fuel resulting from a carbon tax or a cap-and-trade program.

[3] The sources for Table 1 similarly did not contain an estimate of the CO2 emissions of the construction sector, but CBO was able to estimate those emissions at 2 percent of the U.S. total using estimates of industry emissions for 2006 from Department of Commerce, Economics and Statistics Administration, *U.S. Carbon Dioxide Emissions and Intensities Over Time: A Detailed Accounting of Industries, Government and Households* (April 2010), http://go.usa.gov/T8P9; estimates of sector emissions for 2006 and 2010 from Environmental Protection Agency, *Inventory of U.S. Greenhouse Gas Emissions and Sinks: 1990–2010*, EPA 430-R-12-001 (April 2012), p. 2-19, Table 2-14, http://go.usa.gov/T8mj; and industry value-added, output, and price data from the Bureau of Economic Analysis. The estimate of 28 percent for the agricultural, mining, and manufacturing sectors is derived from the numbers in the table and the estimate of 2 percent for the construction sector.

[4] That percentage is based on data from Environmental Protection Agency, *Inventory of U.S. Greenhouse Gas Emissions and Sinks: 1990–2010* (April 2012), Table 2-14, pp. 2-19 to 2-20, http://go.usa.gov/T8mj.

[5] U.N. Framework Convention on Climate Change (UNFCCC), May 9, 1992, 1771 U.N.T.S. 107, http://tinyurl.com/cq9dxdm. Thirty-eight countries signed Annex I to the convention, thereby agreeing to limit their emissions in accordance with the convention. Twenty-three of those countries, generally the advanced industrialized countries, also signed Annex II, thereby agreeing to an additional provision requiring that they provide financial resources to cover the costs incurred by developing countries in Annex I in meeting their obligations under the convention. The Kyoto Protocol to the UNFCCC, Dec. 11, 1997, 2303 U.N.T.S. 148, specified more precise commitments for its signatories, which are listed in Annex B of the protocol. The list of countries is identical to the list in Annex I of the UNFCCC except for Belarus and Turkey, both of which are listed in Annex I but not in Annex B of the Kyoto Protocol.

[6] *Process emissions* are emissions from a production process other than those that result from the consumption of carbon-based fuels. The *value added by an industry* is the value of its output minus the value of the intermediate inputs it purchases from other industries for its production process.

[7] Environmental Protection Agency, *EPA Analysis of the Lieberman-Warner Climate Security Act of 2008, S. 2191 in 110th Congress* (March 14, 2008), http://go.usa.gov/T8VR (PDF, 4 MB). S. 2191 was reported from the Senate Committee on Environment and Public Works on May 20, 2008, but no further action was taken on the bill.

[8] See Liwayway Adkins and others, *Carbon Pricing With Output-Based Subsidies: Impact on U.S. Industries Over Multiple Time Frames*, RFF DP 12-27 (Resources for the Future, June 2012), http://tinyurl.com/lnn2d6s. The study also presents estimates for leakage that would occur if output-based subsidies equivalent to the transition assistance that would be granted under H.R. 2454, the American Clean Energy and Security Act of 2009, were allocated to industries. Those estimates are not presented here, in part because the purpose of this section is to indicate the likely magnitude of leakage if nothing was done to counteract it and in part because the effects of such subsidies on leakage *rates* (as opposed to absolute amounts of leakage) are different for a carbon tax than for a cap-and-trade program.

[9] J.P.M. Sijm and others, *Spillovers of Climate Policy: An Assessment of the Incidence of Carbon Leakage and Induced Technological Change Due to CO2 Abatement Measures*, Report 500036-002 (Netherlands Research Programme on Climate Change, December 2004), pp. 13–14 and 31, http://tinyurl.com/lh3f3pw (PDF, 1.29 MB).

[10] ZhongXiang Zhang, *Competitiveness and Leakage Concerns and Border Carbon Adjustments*, Working Papers in English 2012.080 (Fondazione Eni Enrico Mattei, 2012), p. 3, http://tinyurl.com/clzqb5z.

[11] Jean-Marc Burniaux and Joaquim Oliveira Martins, *Carbon Emission Leakages: A General Equilibrium View*, Working Paper 242 (Organisation for Economic Co-operation and Development, May 2000), http://tinyurl.com/p5tzn4c (PDF, 548 KB).

[12] See Adkins and others, *The Impact on U.S. Industries of Carbon Prices With Output-Based Rebates Over Multiple Time Frames*.

[13] Environmental Protection Agency, *The Effects of H.R. 2454 on International Competitiveness and Emission Leakage in Energy-Intensive Trade-Exposed Industries*, An Interagency Report Responding to a Request From Senators Bayh, Specter, Stabenow, McCaskill, and Brown (December 2, 2009, with corrections on February 23, 2010), http://go.usa.gov/T8pT (PDF, 1.8 MB).

[14] Christoph Böhringer, Edward J. Balistreri, and Thomas F. Rutherford, eds., "The Role of Border Carbon Adjustment in Unilateral Climate Policy: Results from EMF 29," *Energy Economics*, vol. 34, supp. 2 (December 2012), pp. S95–S250, http://tinyurl.com/c65mott.

[15] See, for example, John P. Weyant, ed., *The Costs of the Kyoto Protocol: A Multi-Model Evaluation*, a special issue of *The Energy Journal* (Energy Economics Education Foundation, May 1999), http://tinyurl.com/bowde5c, which contains a number of such studies.

[16] J.P.M. Sijm and others, *Spillovers of Climate Policy: An Assessment of the Incidence of Carbon Leakage and Induced Technological Change Due to CO2 Abatement Measures*, Report 500036-002 (Netherlands Research Programme on Climate Change, December 2004), p. 14, http://tinyurl.com/lh3f3pw (PDF, 1.29 MB).

[17] Jean-Marc Burniaux and Joaquim Oliveira Martins, *Carbon Emission Leakages: A General Equilibrium View*, Working Paper 242 (Organisation for Economic Co-operation and Development, May 2000), http://tinyurl.com/p5tzn4c (PDF, 548 KB).

[18] ZhongXiang Zhang, *Competitiveness and Leakage Concerns and Border Carbon Adjustments*, Working Papers in English 2012.080 (Fondazione Eni Enrico Mattei, 2012), p. 9, http://tinyurl.com/clzqb5z.

[19] Aaditya Mattoo and others, *Reconciling Climate Change and Trade Policy*, Policy Research Working Paper 5123 (World Bank, November 2009), http://tinyurl.com/d9arqqx.

[20] A. Denny Ellerman, Frank J. Convery, and Christian de Perthuis, *Pricing Carbon: The European Union Emissions Trading Scheme* (Cambridge University Press, 2010), http://tinyurl.com/cmhj55v.

[21] Ralf Martin, Laure B. de Preux, and Ulrich J. Wagner, *The Impacts of the Climate Change Levy on Manufacturing: Evidence from Microdata*, Working Paper 17446 (National Bureau of Economic Research, September 2011), www.nber.org/papers/w17446.

[22] Joseph E. Aldy and William A. Pizer, *The Competitiveness Impacts of Climate Change Mitigation Policies*, Working Paper 17705 (National Bureau of Economic Research, December 2011), www.nber.org/papers/w17705.

In: Unintended Effects of Emissions Reduction ISBN: 978-1-63117-902-0
Editor: Kassy Kinsley

Chapter 3

THE EFFECTS OF H.R. 2454 ON INTERNATIONAL COMPETITIVENESS AND EMISSION LEAKAGE IN ENERGY-INTENSIVE TRADE-EXPOSED INDUSTRIES*

U. S. Environmental Protection Agency

I. EXECUTIVE SUMMARY

Proposed greenhouse gas cap-and-trade legislation will play an important role in advancing both the United States' transition to a clean-energy economy and the long-run competitiveness of U.S. manufacturing. Some energy-intensive trade-exposed industries have raised questions, however, about the potential near-term international competitiveness impacts of domestic energy and climate legislation. If the adoption of a domestic cap-and-trade program leads some manufacturing activity and its associated emissions to shift to countries that do not yet have comparable greenhouse gas regulations, along with the economic concerns that this poses, this presents environmental concerns because the resulting "emission leakage" can undermine the environmental effectiveness of a domestic emissions cap. The most effective approach to address these concerns is to ensure significant action by all major

* This is an edited, reformatted and augmented version of an Interagency Report Responding to a Request from Senators Bayh, Specter, Stabenow, McCaskill, and Brown. The report was originally dated December 2, 2009 then revised February 23, 2010.

emitters through ongoing international negotiations. It is important to understand, however, both the scope of potential competitiveness and leakage impacts that could exist prior to securing such action, and also the ability of certain provisions in proposed legislation — particularly the American Clean Energy and Security Act of 2009 (H.R. 2454) — to mitigate those impacts.

This report is a response to a request from Senators Bayh (D-Indiana), Specter (D-Pennsylvania), Stabenow (D-Michigan), McCaskill (D-Missouri), and Brown (D-Ohio). In a September 11, 2009 letter to the White House (see Appendix C), the Senators request "assistance in assessing the ... provisions [addressing energy-intensive trade-exposed industries] in Section 782(e) of ... H.R. 2454." In particular, they seek an analysis that determines which industrial sectors are likely to be eligible for allowance rebates to "trade-vulnerable" industries under H.R. 2454, and that assesses "potential competitiveness impacts to energy intensive and trade exposed manufacturers if the United States adopts a climate change policy...." The Senators further request that the analysis consider the range of measures in H.R. 2454 that may benefit energy-intensive trade-exposed industries, and "identify additional data that, if provided, could improve the assessment of competitiveness impacts."

This report examines patterns of energy use, emissions, and trade across manufacturing sectors to identify vulnerable industries, assesses the potential impacts of proposed legislation on these industries, and estimates how key allowance allocation provisions moderate these impacts on costs, trade flows, and ultimately emission leakage. Given that, on average, energy expenditures account for less than 2 percent of the value of U.S. manufacturing output, the vast majority of U.S. industry will be largely unaffected by proposed legislation. H.R. 2454 establishes thresholds for "presumptive eligibility" for mitigation measures based on an industry's energy intensity, greenhouse gas intensity, and trade intensity. These thresholds offer useful criteria for identifying vulnerable industries. Based on a preliminary assessment using these criteria, 44 of about 500 manufacturing industries would be "presumptively eligible." Almost all of these industries fall within the chemicals, paper, nonmetallic minerals (e.g., cement and glass), or primary metals (e.g., aluminum and steel) sectors. In addition to these industries, the processing subsectors of a few mineral industries are also likely to be deemed eligible. Together, these energy-intensive trade-exposed industries account for 12 percent of total manufacturing output and 6 percent of manufacturing employment (half a percent of total U.S. non-farm employment), but almost half of manufacturing greenhouse gas emissions. In 2006, their emissions were about 730 million metric tons of carbon dioxide equivalent (MMTCO2e).

To understand H.R. 2454's impacts on industry, we draw on prior literature, new analysis of these industries' emission intensities, existing modeling by the Environmental Protection Agency (EPA) and Energy Information Administration (EIA), and new modeling conducted for this report. In assessing impacts on these industries, we focus on the effects of an allowance price of $20 per ton of CO2e. This price was chosen because it is representative of the near- to medium-term allowance prices under H.R. 2454 projected by various studies. For example, Congressional Budget Office and EIA analyses of H.R. 2454 both projected that allowance prices would reach $20 by the middle of the coming decade, while EPA analysis projected that prices would reach $20 somewhat later, after 2020. In modeling impacts on industry, we consider scenarios with and without two key provisions in H.R. 2454 that can reduce competitiveness impacts and leakage, the allocation of allowances to local distribution companies (LDCs) for electricity and the allocations (or rebates) to "trade-vulnerable" industries.

The economic modeling that we conduct examines average impacts on each of five broad energy-intensive trade-exposed industrial sectors: chemicals, plastics and rubber; pulp and paper; nonmetallic minerals; iron and steel; and nonferrous metals (e.g., aluminum). That modeling finds that, at an allowance price of $20 per ton and absent H.R. 2454's allocation provisions, the average increase in production costs experienced by energy-intensive trade-exposed industries within each of these five broad sectors would range from less than 0.5 percent to slightly more than 2.5 percent. In turn, across the five sectors, these increased production costs would lead to average impacts on net imports that, when defined as a percentage of domestic production, generally fall within the same range. However, the modeling also finds that the allocations to LDCs and "trade-vulnerable" industries can eliminate almost all — and, in some cases, potentially more than all — of those cost impacts, as well as the resulting changes in net imports and associated emission leakage. The conclusions regarding the effectiveness of the mitigating allocations appear robust. Analyses of Europe's greenhouse gas cap-and-trade program (the European Union Emission Trading Scheme) that examined key sectors in detail and used a range of allowance prices reached similar conclusions about the ability of output-based allocations to limit a cap-and-trade program's impacts on net imports and associated leakage.

The modeling indicates that, even absent the mitigating allocation measures, total annual emission leakage to unregulated countries associated with a cap-and-trade program's impacts on the international competitiveness of domestic "trade-vulnerable" industries is likely to be only on the order of 10

MMTCO2e. The modeling projects that the vast majority of emission reductions achieved by these industries under a cap-and-trade program will be from reductions in the emission-intensity of their production (*e.g.*, increased energy efficiency, or shifts to lower-emission production methods), rather than from declines in production associated with increased imports from unregulated countries. Importantly, while output-based allocations can essentially eliminate the leakage that is associated with the reduced international competitiveness of domestic industry, if carefully designed, these allocations can do so while preserving incentives for industry to reduce the emission-intensity of its production. With such allocations, leakage associated with impacts on the international competitiveness of domestic industries falls to about one MMTCO2e, or about one percent of the estimated emission reductions from those industries.

There are important caveats to this analysis. Most importantly, absent the mitigating policy measures like those in H.R. 2454, some industries would likely be more significantly affected by climate policy than the average effects reported above. However, even for these industries, well-designed output-based allocations should be equally effective in mitigating the impacts that they would otherwise experience. Also, while the model used for this report has several features that make it preferable to other models in addressing competitiveness and leakage issues, it examines a climate policy's impacts relative to the economy as it was in 2004. In some respects, this feature causes the model to overstate impacts on industry, such as by excluding effects of recent legislation (e.g., the Energy Independence and Security Act of 2007 and the American Recovery and Reinvestment Act of 2009) in increasing the efficiency of manufacturing and growing a clean energy economy. In other respects, this feature causes the model to understate impacts. Further, while the modeling assumes a $20 allowance price and provides a snapshot of impacts under this price, allowance prices are expected to change over time, leading to effects that would also vary over time.

As more information continues to be gathered going forward, this will improve our ability to characterize competitiveness and leakage impacts, as well as the effects of the allocation provisions in H.R. 2454 that can mitigate these impacts. For example, improvements in our understanding of intra-industry variation in the emission-intensity and trade-exposure of firms can help us better understand both the distribution of impacts within an industry, as well as the industry-wide impacts themselves. Some firms in energy-intensive trade-exposed industries that are less emission-intensive than their sector average may experience a net benefit from the imposition of climate policy — when considering both the requirement to submit allowances and also the mitigating allocations — while others that are more emission-intensive than average may still face higher costs after the mitigating allocation measures.

Finally, while the modeling performed in this report assumes straightforward implementation of output-based allocations, in reality, there are challenging implementation issues. For example, for purposes of granting allocations, it will be difficult to define appropriate output metrics across the varied products in some industries (e.g., chemicals). Also, some of the data needed to develop and implement meaningful benchmarks are not currently collected.

In developing this report, the Administration met with a range of stakeholders (see Appendix B), including representatives from affected industries, labor, and non-governmental organizations. We consider this report to be a first step in the Administration's engagement on this issue, and look forward to working with stakeholders to improve both our understanding of these issues and their implications for policy design.

II. INTRODUCTION

In commenting on the House of Representatives' passage of the American Clean Energy and Security Act of 2009 (H.R. 2454), the President declared that such legislation is "...a bold and necessary step that holds the promise of creating new industries and millions of new jobs; decreasing our dangerous dependence on foreign oil; and strictly limiting the release of pollutants that threaten the health of families and communities and the planet itself." As creators, producers, and consumers of clean energy technologies, the U.S. manufacturing sector will play a critical role in achieving our national clean energy and climate change goals. At the same time, while U.S. manufacturers have much to gain in the transition to a clean energy economy, it is important that policies to achieve that transition be carefully designed to avoid unintended adverse impacts on our manufacturing sector.

A greenhouse gas cap-and-trade program can play an important role in advancing the long-run competitiveness of domestic manufacturing sectors as the world transitions to a clean energy economy by providing manufacturers with important incentives for innovation and flexibility as we undergo this transition. In the near term, however, there is a legitimate concern that, absent any mitigating measures, the adoption of a domestic cap-and-trade program could place a limited set of energy-intensive industries at a disadvantage relative to international competitors that do not face similar, contemporaneous regulations. If this leads some manufacturing activity and its associated emissions to shift to countries that do not yet have comparable greenhouse gas regulations, along with the economic concerns that this poses, this also presents environmental concerns because the resulting "emission leakage" can undermine the environmental effectiveness of a domestic emissions cap.[1]

The Administration believes that the most effective approach to address concerns with emission leakage and international competitiveness impacts is to negotiate a new international climate change agreement that ensures that all the major emitters take significant actions to reduce their emissions. Some of our major trading partners already have taken significant action. For example, since 2005, the European Union has had a greenhouse gas cap-and-trade program in place that covers the power sector and major manufacturing emission sources. As the United States moves forward with our energy and climate policies, some trading partners may lag in adopting comparable greenhouse gas regulations. Consequently, some period of time may elapse between when the United States implements a cap-and-trade program and when all of our major trading partners adopt comparable measures. In light of this possibility, the Administration has responded to the request to assess impacts on international competitiveness and leakage during this interim period. Likewise, it is important to evaluate how effectively these impacts can be mitigated by certain provisions of a domestic cap-and-trade program, like those set forth in the American Clean Energy and Security Act of 2009 (H.R. 2454).[2]

This report evaluates a cap-and-trade program's impacts on emission leakage and the international competitiveness of U.S. manufacturing. It begins in Section III by identifying energy-intensive trade-exposed manufacturing industries, and presenting estimates of their emissions. Section IV examines the factors that affect potential leakage in these industries, and Section V reviews the experience of these industries within the European cap-and-trade program. Section VI surveys the provisions in H.R. 2454 and also in the American Recovery and Reinvestment Act of 2009 that assist energy-intensive manufacturing in the transition to a clean energy economy. Section VII uses economic modeling to assess potential impacts of a cap-and trade program on emission leakage and the competitiveness of U.S. manufacturing, and to evaluate the efficacy of particular provisions in H.R. 2454 that have been developed to address concerns about international competitiveness and emission leakage. Finally, Section VIII discusses caveats and next steps.

III. Energy-Intensive Trade-Exposed Industries

Through its requirement that firms acquire and submit allowances to cover their greenhouse gas emissions, a cap-and-trade program like that incorporated in H.R. 2454 will tend to have more significant effects on an industry the more emission-intensive that industry is. While some industries have significant greenhouse gas emissions associated with manufacturing processes that do not involve fuel combustion, most manufacturing emissions are associated with energy use. As a result, the more energy-intensive an industry is, the more emission-intensive it is, and hence the more it will likely be affected by a cap-and-trade program. On the whole, energy expenditures equal only 2 percent of the value of U.S. manufacturing's output (see Figure 1) and three-quarters of all manufacturing output is from industries with energy expenditures below 2 percent of the value of their output. Thus, the vast majority of U.S. industry will be relatively unaffected by a greenhouse gas cap-and-trade program. Industries whose energy expenditures exceed five percent of the value of their output — a threshold given significance in H.R. 2454, as described below — account for only one-tenth of the value of U.S. manufacturing output, and less than two percent of U.S. gross domestic product in 2007. Thus, while concerns have been expressed about a cap-and-trade program's impacts on U.S. manufacturing, it is important to recognize that these concerns apply only to a small subset of manufacturers, and thereby call for narrowly and carefully targeted policies.

H.R. 2454 establishes specific criteria that industries must meet to be eligible for the provisions that are intended to address emission leakage associated with impacts on international competitiveness. Using the best and most recent publicly available data, the remainder of this section identifies the specific energy-intensive trade-exposed industries that would likely be deemed eligible for these provisions, and quantifies their recent emissions.[3]

A. Identifying Energy-Intensive Trade-Exposed Industries

In the course of establishing provisions to address emission leakage associated with the international competitiveness impacts of a domestic cap-and-trade program, H.R. 2454 establishes criteria for identifying energy-intensive trade-exposed industries.[4]

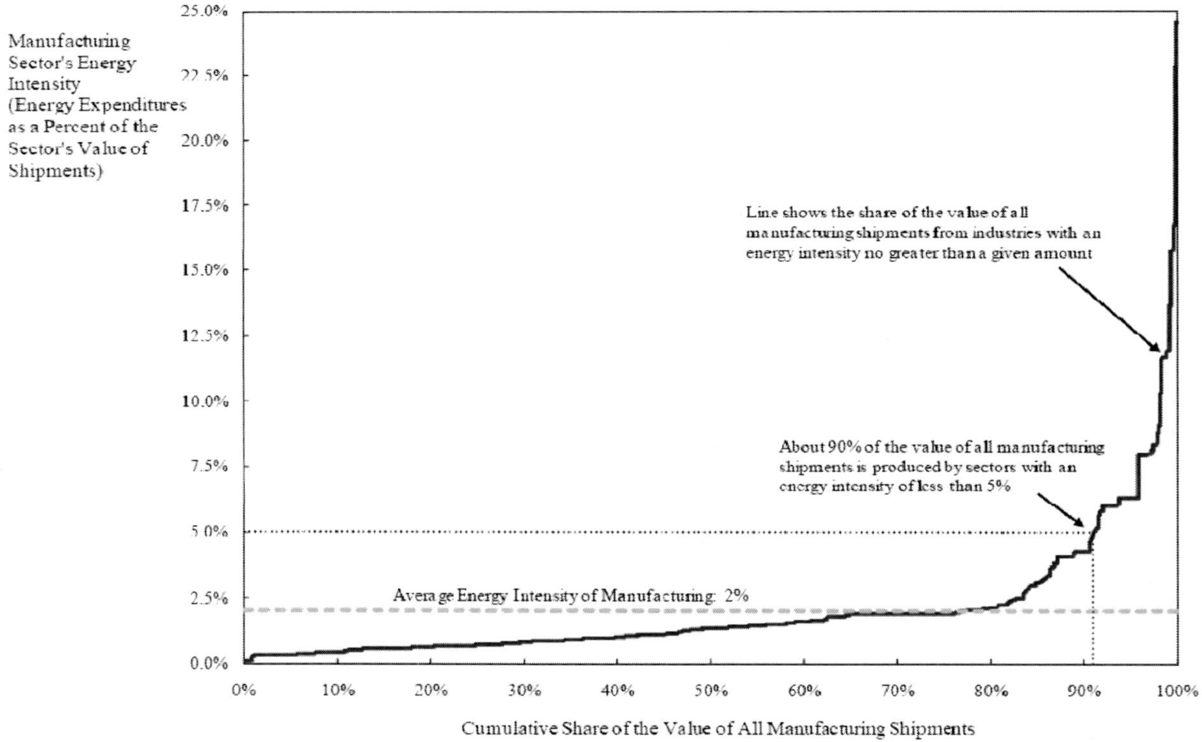

Note: This chart does not account for 14 (of 471) manufacturing sectors for which data on energy expenditures were withheld in the Economic Census to maintain confidentiality of individual companies' data. However, these sectors account for only 2 percent of the value of all manufacturing shipments. Source: 2007 Economic Census.

Figure 1. Energy Intensity of U.S. Manufacturing Sectors in 2007.

Specifically, H.R. 2454 considers an industry to be "presumptively eligible" for emission allowance allocations (or "rebates") to "trade-vulnerable" industries if the industry's energy intensity or its greenhouse gas intensity is at least 5 percent, and its trade intensity is at least 15 percent.[5]

In addition, H.R. 2454 considers an industry to be "presumptively eligible" if its energy or greenhouse gas intensity is at least 20 percent, regardless of its trade intensity. H.R. 2454 stipulates the specific data sources that should be relied on in assessing industry eligibility; these include the Census Bureau's Annual Survey of Manufactures and Economic Census, the Energy Information Administration's (EIA) Manufacturing Energy Consumption Survey, and data from the United States International Trade Commission. The bill also requires that, to the extent feasible, eligibility assessments should be conducted at the most disaggregated level for which the necessary public data are available — the six-digit industry classification under the North American Industry Classification System (NAICS).[6]

The U.S. Environmental Protection Agency (EPA) has previously developed a preliminary assessment of the sectors that would likely be deemed "presumptively eligible" for allowance allocations to "trade-vulnerable" industries under H.R. 2454. However, the recent release of updates to key data sources has allowed EPA to revise that preliminary assessment.[7]

Table 1 presents the set of six-digit industries that would likely be deemed "presumptively eligible" for allocations under H.R. 2454 based on EPA's updated preliminary assessment. For each industry, Table 1 also presents estimates of the industry's emissions as well as relevant economic characteristics, such as employment, output, energy intensity, greenhouse gas intensity, and trade intensity. The final determination of "presumptively eligible" industries would be made in a formal EPA rulemaking upon enactment of legislation. Moreover, emission estimates in Table 1 are intended to give a sense for the overall scale of the industries' recent emissions and would not be used for eligibility determinations or allocations.

According to the preliminary assessment of the nearly 500 six-digit manufacturing industries, 44 would be deemed "presumptively eligible" for allowance rebates under H.R. 2454. Of these, 12 are in the chemicals sector, 4 are in the paper sector, 13 are in the nonmetallic minerals sector (e.g., cement and glass manufacturers), and 8 are in the primary metals sector (e.g., aluminum and steel manufacturers).

Table 1. Preliminary Assessment of Industries Likely to Be Deemed "Presumptively Eligible" for Allowance Rebates under Section 401 of H.R. 2454[1]

	2007 Economic Statistics[2]			Eligibility Determination		2006 Emissions (MMTCO2e)5				GHG Intensity
NAICS Code and Title	Value of Shipments ($1,000)	Employees	Establishments	Energy Intensity[3]	Trade Intensity[4]	Direct Combustion Emissions	Process Emissions	Indirect Electricity Emissions	Total Emissions	at $20 per Ton of CO2e[6]
311213: Malt manufacturing	786,571	1,022	25	9%	30%	0.8	0.0	0.3	1.0	2.6%
311221: Wet Corn Milling	12,117,145	8,448	64	10%	19%	12.1	0.0	4.3	16.4	3.3%
311613: Rendering and Meat Byproduct Processing	3,563,862	9,355	228	8%	28%	3.5	0.0	0.7	4.2	2.3%
313111: Yarn Spinning Mills	5,011,244	24,750	249	5%	32%	0.4	0.0	2.9	3.3	1.3%
314992: Tire Cord and Tire Fabric Mills	1,069,765	3,577	22	5%	35%	0.2	0.0	0.6	0.8	1.4%
321219: Reconstituted Wood Product Manufacturing	6,896,468	20,426	262	7%	38%	2.7	0.0	4.1	6.8	1.7%
322110: Pulp Mills	5,027,395	7,268	39	8%	90%	2.7	0.0	1.1	3.8	1.8%
322121: Paper (except Newsprint) Mills[7]	46,291,440	75,921	241	8%	17%	29.3	0.0	14.6	44.0	1.9%
322122: Newsprint Mills	3,440,645	4,917	21	16%	68%	2.7	0.0	6.0	8.7	5.1%
322130: Paperboard Mills[7]	25,354,745	36,641	187	12%	25%	19.3	0.0	14.0	33.3	2.9%
325110: Petrochemical Manufacturing	77,661,772	9,257	56	6%	17%	48.5	0.7	3.0	52.2	1.6%
325131: Inorganic Dye and Pigment Manufacturing	5,689,517	7,606	96	6%	43%	2.5	0.0	1.8	4.4	1.5%
325181: Alkalies and Chlorine Manufacturing (incl soda ash beneficiation)	6,370,780	6,364	49	24%	29%	7.8	4.2	3.9	16.0	5.0%
325182: Carbon Black Manufacturing	1,487,557	1,591	32	7%	26%	0.7	3.0	0.4	4.1	5.4%

NAICS Code and Title	2007 Economic Statistics[2]			Eligibility Determination		2006 Emissions (MMTCO2e)5				GHG Intensity at $20 per Ton of CO2e[6]
	Value of Shipments ($1,000)	Employees	Establishments	Energy Intensity[3]	Trade Intensity[4]	Direct Combustion Emissions	Process Emissions	Indirect Electricity Emissions	Total Emissions	
325188: All Other Basic Inorganic Chemical Manufacturing	22,828,592	35,801	631	8%	58%	7.3	5.1	15.4	27.8	2.9%
325192: Cyclic Crude and Intermediate Manufacturing	5,947,517	3,006	31	5%	102%	1.8	0.0	1.7	3.5	1.2%
325199: All Other Basic Organic Chemical Manufacturing	81,997,462	70,602	818	6%	49%	34.4	5.9	13.7	54.0	1.3%
325211: Plastics Material and Resin Manufacturing	85,231,585	71,216	1,059	5%	38%	28.1	0.0	12.2	40.3	1.0%
325212: Synthetic Rubber Manufacturing	8,253,660	9,794	152	5%	57%	2.4	0.0	1.1	3.6	1.0%
325221: Cellulosic Organic Fiber Manufacturing	925,820	1,353	15	7%	90%	0.7	0.0	0.2	0.9	1.9%
325222: Noncellulosic Organic Fiber Manufacturing	5,963,293	14,684	109	5%	40%	4.4	0.0	3.2	7.6	2.2%
325311: Nitrogenous Fertilizer Manufacturing	5,524,151	3,920	156	14%	83%	6.4	30.5	1.5	38.4	18.5%
327111: Vitreous China Plumbing Fixture and China and Earthenware Bathroom Accessories	867,553	4,825	30	5%	60%	0.7	0.0	0.1	0.9	1.8%
327112: Vitreous China, Fine Earthenware, and Other Pottery Product Manufacturing	783,594	8,774	664	5%	94%	0.5	0.0	0.1	0.7	1.4%
327113: Porcelain Electrical Supply Manufacturing	737,282	4,465	113	5%	41%	0.4	0.0	0.3	0.7	1.3%
327122: Ceramic Wall and Floor Tile Manufacturing	1,126,093	6,272	183	7%	69%	1.1	0.0	0.3	1.4	2.4%

Table 1. (Continued)

NAICS Code and Title	2007 Economic Statistics[2]			Eligibility Determination		2006 Emissions (MMTCO2e)5				GHG Intensity at \$20 per Ton of CO2e[6]
	Value of Shipments (\$1,000)	Employees	Establishments	Energy Intensity[3]	Trade Intensity[4]	Direct Combustion Emissions	Process Emissions	Indirect Electricity Emissions	Total Emissions	
327123: Other Structural Clay Product Manufacturing	243,009	1,650	54	10%	27%	0.4	0.0	0.0	0.4	3.5%
327125: Nonclay Refractory Manufacturing	1,372,439	5,338	101	6%	46%	0.9	0.0	0.3	1.1	1.7%
327211: Flat Glass Manufacturing	3,420,860	10,991	93	16%	51%	2.9	0.1	1.1	4.2	2.5%
327212: Other Pressed and Blown Glass and Glassware Manufacturing	4,316,979	21,189	524	11%	58%	4.1	0.1	3.2	7.4	3.7%
327213: Glass Container Manufacturing	4,899,025	14,928	74	14%	21%	2.7	0.1	2.4	5.3	2.4%
327310: Cement Manufacturing	10,619,945	17,749	302	15%	19%	31.0	46.6	7.7	85.3	15.9%
327410: Lime Manufacturing	1,875,567	4,369	83	23%	4%	10.4	15.1	0.9	26.4	33.0%
327992: Ground or Treated Mineral and Earth Manufacturing	2,826,839	6,497	258	9%	17%	3.1	0.0	1.0	4.1	3.0%
327993: Mineral Wool Manufacturing	6,147,076	18,891	307	8%	18%	2.2	0.0	2.6	4.8	1.5%
331111: Iron and Steel Mills[8]	102,186,442	114,315	743	6%	33%	101.3	0.0	32.8	134.1	2.6%
331112: Electrometallurgical Ferroalloy Product Manufacturing	1,319,541	2,144	20	11%	77%	0.5	1.5	2.1	4.0	6.1%
331210: Iron and Steel Pipe and Tube Manufacturing from Purchased Steel[8]	8,637,314	17,408	153	2%	N/A	1.4	0.0	1.2	2.6	0.6%
331311: Alumina Refining	1,337,014	1,611	16	21%	70%	2.3	6.7	24.3	33.4	9.0%
331312: Primary Aluminum Production	6,657,285	9,355	54	22%	64%	Emissions and GHG Intensity Reflected in Alumina Refining				

NAICS Code and Title	2007 Economic Statistics[2]			Eligibility Determination		2006 Emissions (MMTCO2e)5				GHG Intensity at $20 per Ton of CO2e[6]
	Value of Shipments ($1,000)	Employees	Establishments	Energy Intensity[3]	Trade Intensity[4]	Direct Combustion Emissions	Process Emissions	Indirect Electricity Emissions	Total Emissions	
331411: Primary Smelting and Refining of Copper[9]	8,247,767	1,771	13	2%	55%	1.0	0.0	0.3	1.3	0.3%
331419: Primary Smelting and Refining of Nonferrous Metal (except Copper and Aluminum)	5,987,185	8,067	183	7%	135%	1.1	0.8	3.8	5.7	1.9%
331511: Iron Foundries	11,795,934	51,503	470	6%	15%	4.0	0.0	5.4	9.4	1.6%
335991: Carbon and Graphite Product Manufacturing	2,795,262	8,666	143	6%	52%	0.7	0.0	1.0	1.7	1.4%
212210: Iron Ore Mining	2,955,254	4,989	22	11%	38%	6.3	0.0	4.4	10.7	7.3%
212234: Copper Ore and Nickel Ore Mining[9]	8,985,692	10,384	31	6%	12%	6.7	0.0	3.2	9.9	2.2%
TOTALS	618,581,937	783,670	9,176			405	120	205	730	

Notes:

[1] This table updates previous EPA analysis (released June 10, 2009) with recently released data, including the 2007 Economic Census and 2006 Manufacturing Energy Consumption Survey. This does not represent a final EPA determination and the emissions estimates presented here will not be used to make any allocation determinations. The methodology is detailed in a separate EPA memorandum available upon request.

[2] While these statistics reflect 2007 data, the energy, greenhouse gas, and trade intensities presented to the right are typically based on data from earlier years, as required in H.R. 2454 and described in the above-referenced memorandum.

[3] Energy intensity is equal to a sector's energy expenditures divided by its value of shipments. The specific calculations and sources of relevant data are detailed in the above-referenced memorandum.

[4] Trade intensity is equal to the sum of the value of a sector's imports and exports, divided by the sum of its value of shipments and imports. The specific calculations and sources of relevant data are detailed in the above-referenced memorandum.

[5] The Energy Information Administration's 2006 Manufacturing Energy Consumption Survey was used to estimate direct combustion emissions and indirect electricity emissions from 25 industries that account for 90% of the total estimated direct combustion and

indirect electricity emissions of the presumptively eligible industries. The remaining emissions estimates are based on EPA analysis that relies on several data sources. All process emissions estimates are derived from EPA's *Inventory of U.S. Greenhouse Gas Emissions and Sinks: 1990-2007*. Given data limitations, emissions estimates for 10 sectors (with total estimated emissions of 43 MMTCO2e) reflect estimates of 2007, rather than 2006, emissions. The above-referenced memorandum provides additional details regarding emissions estimation methods.

[6] As called for in H.R. 2454, greenhouse gas intensity is calculated by monetizing an industry's emissions at a price of $20 per ton of carbon dioxide equivalent, and dividing this value by the industry's value of shipments.

[7] The U.S. Census Bureau recently adjusted the classifications that it employs in assigning imports and exports to the paper and paperboard industries, and will report revised data beginning with 2009. The trade intensities reported here reflect EPA's preliminary analysis of how this adjustment would affect reported trade data in prior years. The above-referenced memorandum provides additional details regarding the methodology employed to determine the trade intensity of these sectors.

[8] Iron and steel mill process emissions are included in the "Direct Combustion" estimate due to the nature of the data collection and reporting by the Manufacturing Energy Consumption Survey. The above-referenced EPA memorandum provides additional details explaining this categorization.

[9] On their own, these sectors do not meet either the energy or trade intensity thresholds specified in the bill, but are expected to be eligible based on other language in the bill. The above-referenced memorandum provides additional details regarding eligibility determination for these sectors.

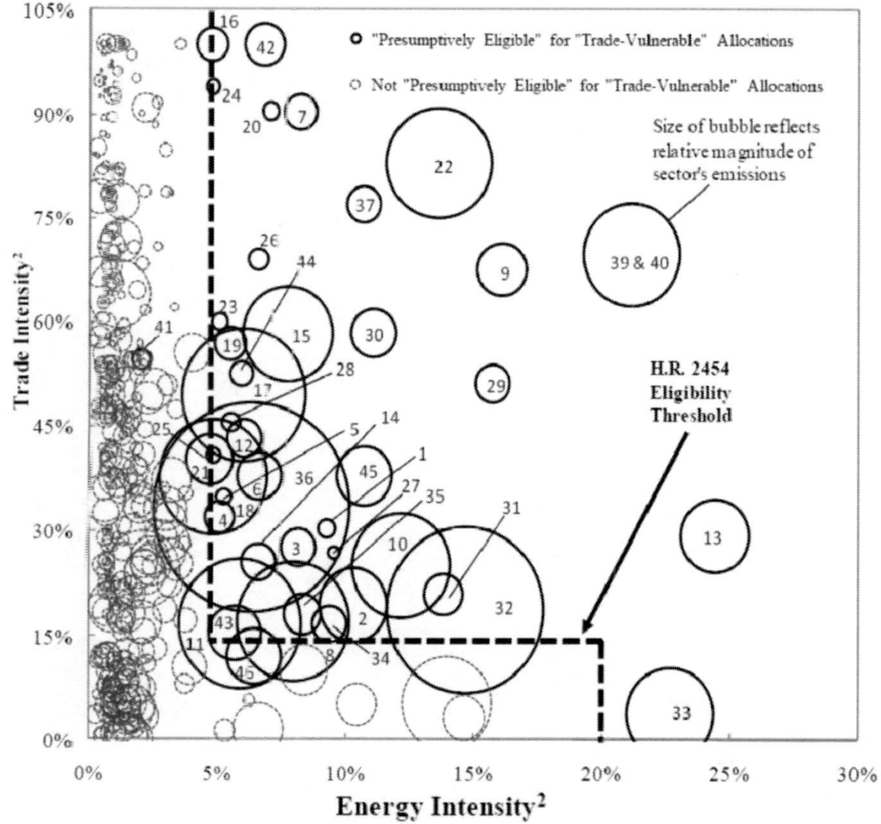

Figure 2. (Continued).

Notes:

1) Petroleum refining is not depicted because it is explicitly excluded from H.R. 2454's allocations to "trade-vulnerable" industries. Also, 91 other sectors, with 130 MMTCO2e of emissions, are not depicted due to lack oftrade intensity data. One of these, iron and steel pipe and tube manufacturing from purchased steel (331210; 2.6 MMTCO2e) is expected to be eligible based on language in the bill. Four others meet the energy-intensity threshold, each with 2 to 3 MMTCO2e of emissions: beet sugar manufacturing, broadwoven fabric finishing mills, steel foundries (except investment), and metal heat treating. Twelve sectors with a calculated trade intensity greater than 100% are depicted here with an intensity of 100% (the maximum possible intensity). The two copper sectors (212234 and 331411) do not meet the energy or trade intensity thresholds specified in H.R. 2454 but are expected to be eligible based on other language in the bill.
2) Energy intensity and trade intensity measures are as defined in H.R. 2454 and elsewhere in this report. Source: EPA analysis.

1) Malt Manufacturing (311213)
2) Wet Corn Milling (311221)
3) Rendering and Meat Byproduct Processing (311613)
4) Yarn Spinning Mills (313111)
5) Tire Cord and Tire Fabric Mills (314992)
6) Reconstituted Wood Product Manufacturing (321219)
7) Pulp Mills (322110)
8) Paper (except Newsprint) Mills (322121)
9) Newsprint Mills (322122)
10) Paperboard Mills (322130)
11) Petrochemical Manufacturing (325110)
12) Inorganic Dye and Pigment Manufacturing (325131)
13) Alkalies and Chlorine Manufacturing (325181)
14) Carbon Black Manufacturing (325182)
15) All Other Basic Inorganic Chemical Mfg. (325188)
16) Cyclic Crude and Intermediate Manufacturing (325192)
17) All Other Basic Organic Chemical Mfg. (325199)
18) Plastics Material and Resin Manufacturing (325211)
19) Synthetic Rubber Manufacturing (325212)
20) Cellulosic Organic Fiber Manufacturing (325221)
21) Noncellulosic Organic Fiber Manufacturing (325222)

22) Nitrogenous Fertilizer Manufacturing (325311)
23) Vitr. China Plumbing Fixture and Other Mfg. (327111)
24) Vitreous China and Other Pottery Mfg. (327112)
25) Porcelain Electrical Supply Manufacturing (327113)
26) Ceramic Wall and Floor Tile Manufacturing (327122)
27) Other Structural Clay Product Manufacturing (327123)
28) Nonclay Refractory Manufacturing (327125)
29) Flat Glass Manufacturing (327211)
30) Other Pressed/Blown Glass and Glsswr. Mfg. (327212)
31) Glass Container Manufacturing (327213)
32) Cement Manufacturing (327310)
33) Lime Manufacturing (327410)
34) Ground or Treated Mineral and Earth Mfg. (327992)
35) Mineral Wool Manufacturing (327993)
36) Iron and Steel Mills (331111)
37) Electrometalurgical Ferroalloy Product Mfg. (331112)
38) Iron/Steel Pipe/Tube Mfg. from Purchsd. Steel (331210)
39) Alumina Refining (331311)
40) Primary Aluminum Production (331312)
41) Primary Smelting and Refining of Copper (331411)
42) Smltg./Rfg. of Nonfrs. Mtl. (ex. Cpr. and Almn.) (331419)
43) Iron Foundries (331511)
44) Carbon and Graphite Product Manufacturing (335991)
45) Iron Ore Mining (212210)
46) Copper Ore and Nickel Ore Mining (212234)

Figure 2. Energy Intensity, Trade Intensity, and Emissions of U.S. Manufacturing Sectors at the Six-Digit NAICS Code Level.[1]

Many of these sectors are at or near the beginning of the value chain, and provide the basic materials needed for manufacturing advanced technologies. In addition to these 44 industries, the processing subsectors of a few mineral industries are also likely to be deemed “presumptively eligible.” In total, in 2007, the “presumptively eligible” industries accounted for 12 percent of total manufacturing output and employed about 780,000 workers, or about 6 percent of manufacturing employment and half a percent of total U.S. non-farm employment. As Figure 2 indicates, most industrial sectors have energy intensities of less than 5 percent, and will therefore have minimal direct exposure to a climate policy’s economic impacts.

B. Emissions of Energy-Intensive Trade-Exposed Industries

While accounting for a relatively small share of manufacturing output and employment, the “presumptively eligible” industries’ greenhouse gas emissions were about 730 million metric tons of carbon dioxide equivalent (MMTCO2e) in 2006 (the most recent year for which key data sources are available), or about half of U.S. manufacturing greenhouse gas emissions and 10 percent of total U.S. emissions in that year.[8] In turn, relatively few industries account for the bulk of the “presumptively eligible” industries’ emissions. The top five industries on an emissions basis (iron and steel, cement, other basic organic chemicals, petrochemicals, and paper mills) account for half of the “presumptively eligible” industries’ emissions, and the top ten industries account for three-quarters of those emissions.

The emissions estimates presented in Table 1 are based on data from the Manufacturing Energy Consumption Survey (MECS), the Annual Survey of Manufactures, the Economic Census, the Environmental Protection Agency, and the Bureau of Economic Analysis.[9] While these data provide the best available estimates of recent emissions, there is considerable uncertainty as to what the emissions of the “presumptively eligible” industries will be in 2014, when they would first be directly covered by the cap-and-trade program proposed in H.R. 2454. Over time, the emission intensity of these industries’ production processes is expected to continue to decline as new technology emerges and investments in energy efficiency continue to be made (see Figure 7) putting downward pressure on emissions. At the same time, increases in manufacturing activity in key sectors could have a countervailing effect on emissions.

H.R. 2454 allows for the administrative determination of eligibility for additional sectors based on changes in international trade patterns and based on “individual showings,” whereupon industries and subsectors of industries can petition for eligibility.[10] Figure 3 presents the emissions of all manufacturing industries, aggregating those not deemed “presumptively eligible” into various categories of trade and energy intensity. Figure 3 offers some insight into the quantity of emissions associated with certain industries or subsectors that might ultimately be deemed eligible for allowance rebates. For example, there are about 19 MMTCO2e of emissions associated with those industries that meet the 5 percent energy intensity threshold but have trade intensities that are either between 10 and 15 percent, or are not reported in government trade databases. Two examples of these industries are Steel Foundries and Broadwoven Fabric Mills. On the other hand, it is noteworthy that the vast majority of manufacturing’s remaining emissions are in industries with energy intensities well below the 5 percent threshold. Therefore, it is unlikely that subsequent eligibility determinations, such as those arising from “individual showings”, would dramatically affect the scope of emissions associated with those industries that are deemed eligible for allowance allocations under H.R. 2454.

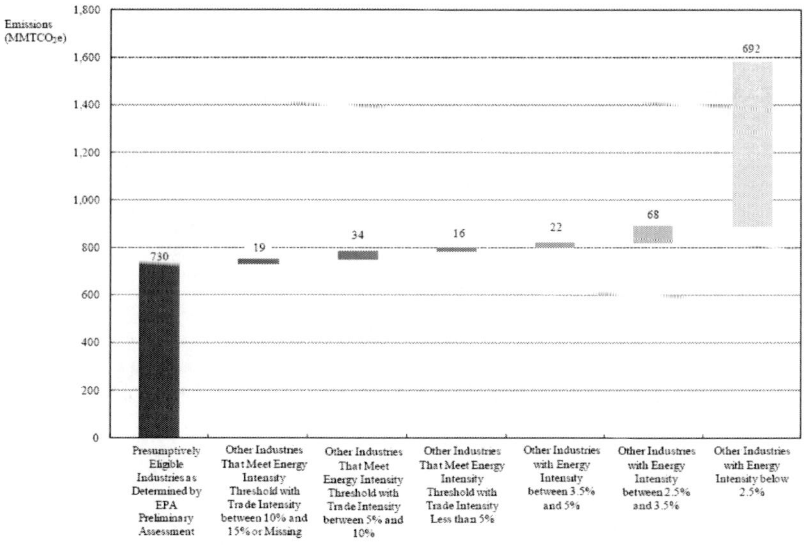

Source: EPA analysis.

Figure 3. Emissions of “Presumptively Eligible” Industries and of Remaining Six-Digit Manufacturing Industries with Various Energy and Trade Intensities.

Figure 4 differs from Figure 3 by presenting the number of employees, rather than emissions, in all manufacturing industries, again aggregated into the "presumptively eligible" industries and into various categories of trade and energy intensity for those industries not deemed presumptively eligible. The U.S. manufacturing sector employed over 13 million people in 2007, representing about 10 percent of total nonfarm employment. This figure shows that the overwhelming majority — nearly 95 percent — of employees in the manufacturing sector fall outside the "presumptively eligible" industries. Indeed, 88 percent of manufacturing employees work in industries with energy intensities below 2.5 percent.

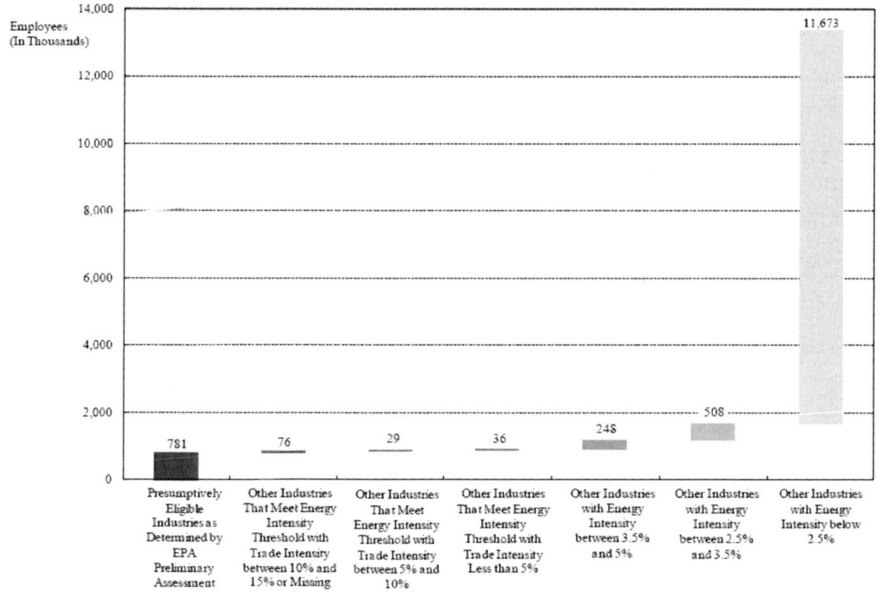

Source: EPA analysis and 2007 Economic Census.

Figure 4. 2007 Employment in "Presumptively Eligible" Industries and in Remaining Six-Digit Manufacturing Industries with Various Energy and Trade Intensities.

IV. Factors Affecting Potential Leakage in Energy-Intensive Trade-Exposed Industries

Through the requirement that firms surrender emission allowances to cover every ton of their emissions, a cap-and-trade program introduces important incentives for firms to switch to less emission-intensive production methods. At the same time, the allowance requirement can also increase a firm's production costs, both through the cost of switching to less emission-intensive processes and through the cost of having to surrender valuable allowances to cover the remaining emissions associated with producing the firm's goods. This latter effect is important in minimizing the cost of meeting national emission targets because it ensures that the resulting market prices of emission-intensive products provide consumers with appropriate incentives to shift to less emission-intensive products and to use emission-intensive products more efficiently.

Yet, in the case of energy-intensive trade-exposed industries, in addition to encouraging consumers to use emission-intensive products more efficiently, changes in domestic production costs of emission-intensive goods can cause shifts in manufacturing activity and associated emissions to countries that have not yet implemented comparable climate policies. These shifts can occur through both increases in imports to meet domestic demand for emission-intensive goods and reductions in exports from the United States to other markets.

Absent any mitigating policy measures like those proposed in H.R. 2454, the extent to which a domestic cap-and-trade program would lead to economic and emission leakage in a particular industry will depend on how much such a program would affect the industry's domestic production costs, and on the extent to which a given change in those costs would lead to increased imports or reduced exports. Both considerations are addressed in turn below.

A. Impacts of a Cap-and-Trade Program on Industries' Production Costs

In the absence of any mitigating policy measures like those proposed in H.R. 2454, the primary determinants of a cap-and-trade program's effect on an industry's production costs are the emission allowance price, the emission-intensity of the industry's production — taking into account both its direct emissions and its indirect emissions from its electricity use — and the industry's ability to shift to less emission-intensive production methods. The higher the allowance price, the more emission-intensive the industry's production, and the less able an industry is to shift to less emission-intensive production methods, the greater will be the cap-and trade program's impact on the industry's production costs.

In addition to the above factors, through its broader effects on the economy, a cap-andtrade program can have other indirect effects on industries' production costs. In particular, a cap-and-trade program can affect the cost of various raw materials (e.g., steel) that an industry uses. Through its effect on the demand for various fuels, such as coal and natural gas, a cap-and-trade program can also affect the underlying price of those fuels (i.e., the price excluding any allowance costs) beyond the effect directly arising from the emission allowance requirement.

A primary concern of many in industry, particularly those that use natural gas as a feedstock, is a cap-and-trade program's effect on the demand for and underlying price of natural gas.[11] However, it is important to recognize that a cap-and-trade program will not necessarily increase the underlying price of natural gas. In fact, it could reduce the underlying price of natural gas. A cap-and-trade program will reduce demand for natural gas in some applications by making energy efficiency improvements and renewable energy more economically attractive. At the same time, demand for natural gas could increase in other applications where natural gas use becomes more attractive than using coal or oil. Reflecting these countervailing effects, over the years, economic modeling has reached varying conclusions about whether and to what extent a cap-and-trade program would increase or reduce the total demand for and price of natural gas.

Most current modeling projects relatively small impacts on the underlying price of natural gas, with a majority of studies projecting a small decline. For example, in its core scenario analysis, EPA's modeling of H.R. 2454 projects that the underlying price (without allowance value) of natural gas decreases 0.6 percent in 2015 and 1.5 percent in 2020 relative to business-as-usual (BAU) levels.[12] Across all scenarios that EPA examined, using different assumptions about the deployment of energy technologies and implementation of energy efficiency provisions, EPA's projections of changes in the underlying delivered price of natural gas (excluding allowance costs) during this time frame range from a 1.5 percent reduction to a 0.4 percent increase relative to BAU. In sum, EPA's analysis generally finds that the cap-and trade program, the Combined Efficiency and Renewable Electricity Standard, and additional energy efficiency provisions in H.R. 2454 will lead to an expansion of renewable energy and energy efficiency that is expected to ease demand for natural gas and lower its underlying price. The core scenario in the EIA analysis of H.R. 2454 projected that wellhead natural gas prices would increase by 3.9 percent in 2015 and fall by about 5.6 percent in 2020 relative to BAU.[13] Across all 11 scenarios that EIA examined, in all but one, it projected increases in wellhead natural gas prices ranging from 0.5 and 4.7 percent in 2015, and decreases in prices ranging from 1.3 to 8.6 percent in 2020.[14] In analyzing H.R. 2454, the climate modeling group at the Massachusetts Institute of Technology projected modest reductions in underlying natural gas prices (excluding allowance costs) relative to BAU levels in both 2015 and 2020.[15]

While a cap-and-trade program may increase or decrease the demand for and the underlying price of natural gas, because coal and petroleum products are more carbon-intensive fuels, a cap-and-trade program is unambiguously expected to reduce the demand for and underlying price of those fuels. For example, in its core scenario, EPA's modeling results from ADAGE show that the underlying prices for petroleum products and coal decrease by 0.3 percent and 2.2 percent respectively in 2015 relative to BAU projections. Both EIA and MIT also project reductions in the underlying prices of these fuels.

Despite these considerations, the allowance price and emission intensity remain the dominant factors determining cost impacts. In the criteria that it establishes for determining industries that are "presumptively eligible" for allocations to "trade-vulnerable" industries, H.R. 2454 develops a greenhouse gas (GHG) intensity metric that captures these two factors. In particular, the H.R. 2454 GHG intensity measure is calculated by multiplying a $20 allowance price by an industry's direct emissions and indirect emissions associated with electricity consumption, and then dividing this value by the value of the industry's output. Figure 5 presents each manufacturing sector's GHG intensity metric as calculated under H.R. 2454.

Under a cap-and-trade program like that in H.R. 2454, the actual increase in production costs that manufacturers would incur could be less than — potentially significantly less than — what is depicted in Figure 5, even before accounting for the allocations that are specifically designed to mitigate those impacts. There are at least two reasons why this is the case.

First, while H.R. 2454's GHG intensity metric reflects the impacts of a $20 allowance price, EPA's analysis of H.R. 2454 produced an allowance price projection that, in 2005 dollars, begins at $13 in 2015, and does not reach $20 per ton until 2025, which would imply proportionately lower near-term impacts on production costs of energy-intensive trade-exposed industries. Second, Figure 5 is generally based on 2006 data and, going forward, if industries continue their historical trend, they will realize significant reductions in the energy- and emission-intensity of their production, reducing the impact of the allowance requirement on their production costs.

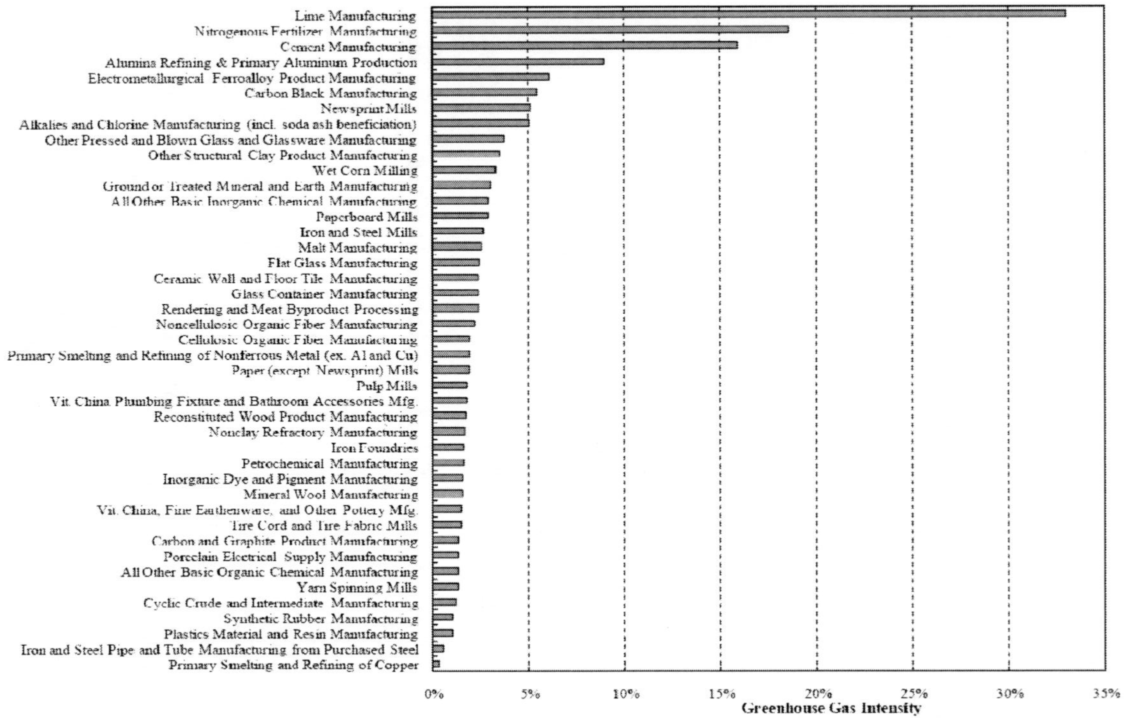

Note: The H.R. 2454 greenhouse gas intensity measure is calculated by multiplying a sector's direct emissions and indirect emissions associated with electricity consumption by an allowance price of $20 per ton, and dividing by the sector's value of shipments.

Source: EPA anaylsis.

Figure 5. H.R. 2454's Greenhouse Gas Intensity Measure for the “Presumptively Eligible” Manufacturing Industries.

Speaking to the latter, Figure 6 shows changes in the energy intensity of production in five energy-intensive sectors from 1998 to 2006. During this period, energy intensity fell between 10 and 35 percent in these sectors. Despite these past reductions in emission- and energy-intensity, significant opportunities for further reductions remain, as state-of-the-art energy management practices evolve, new technologies become commercially available, and existing technologies are more widely deployed.[16] Recognizing these opportunities, many companies and industry groups have set aggressive forward-looking goals as part of voluntary initiatives such as EPA's ENERGY STAR, Climate Leaders, and industrial non-CO_2 programs, and the Department of Energy's Climate VISION program. As Figure 7 depicts, the Energy Information Administration projects that the emission intensity of some of the key energy-intensive sectors will fall over the coming decade even in the absence of a cap-and-trade program, and would fall even more under H.R. 2454. The primary drivers of these projected improvements include: the on-going shift from the use of virgin raw materials to the less emission-intensive use of recycled raw materials, particularly in the aluminum, paper, and iron and steel industries; process improvements, such as the shift from a wet to a dry process in the cement industry; the penetration of more energy efficient systems and practices (*e.g.*, combined heat and power, efficient motors, waste heat recovery); and the increased use of lower carbon fuels like natural gas and renewables. Overall, taking into account both changes in output and emission intensity, EIA projects that the total energy-related CO_2 emissions of the six sectors depicted in Figure 7 would decline nearly 20 percent from 2006 to 2020 under business-as-usual circumstances, and nearly 30 percent under H.R. 2454.

At a $20 allowance price, a cap-and-trade program would increase the production costs of most energy-intensive industries' by no more than a few percentage points even in the absence of any mitigating policy measures. As Figure 5 depicts, however, a small number of these industries would experience more substantial impacts on their production costs absent any mitigating policy measures. Moreover, within the industry categories that are depicted here, there can be significant variation in subsectors' or individual facilities' energy intensity (and hence emission intensity) of production. For example, among iron and steel mills, integrated mills employing basic oxygen furnaces have been found to be nearly three times more energy intensive than electric arc furnaces, another common method of steel production.[17,18]

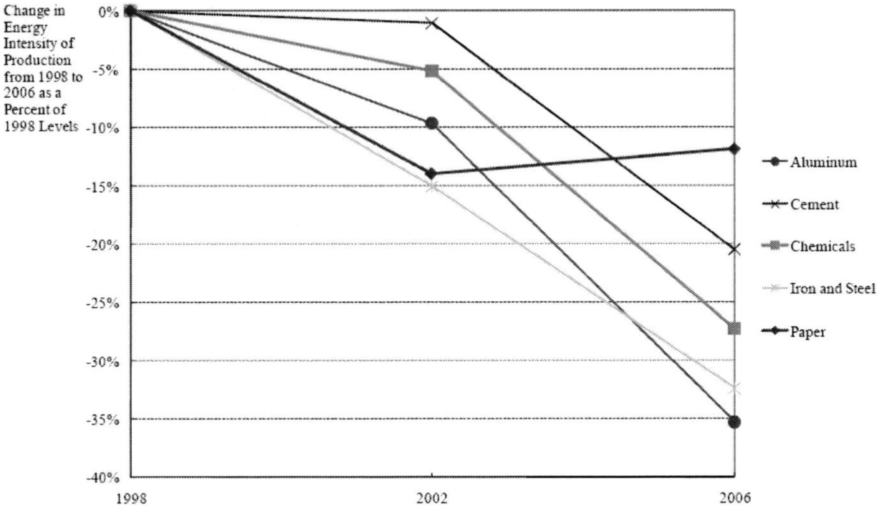

Note: Energy intensity of production is equal to a sector's direct and indirect energy use divided by the real value of its shipments, using industry-specific adjustments for inflation.

Source: EIA analysis of the Manufacturing Energy Consumption Survey.

Figure 6. Changes in the Energy Intensity of Production from 1998 to 2006 in Five Energy-Intensive Industrial Sectors.

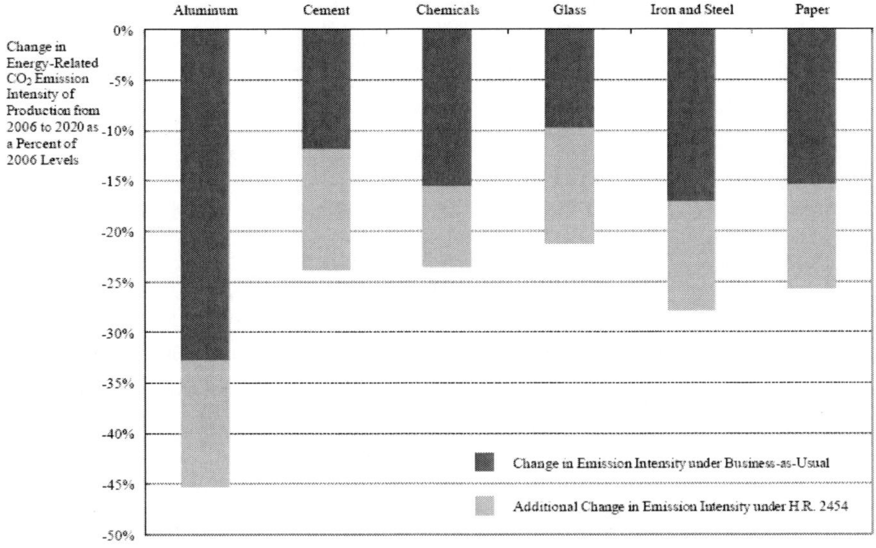

Figure 7. (Continued).

Note: Emission intensity of production is equal to a sector's energy-related direct and indirect carbon dioxide emissions divided by its value of shipments in real 2000 dollars. Reductions in emission intensity result from reductions in both on-site energy use and the emission-intensity of purchased electricity. In addition to reflecting the effect of the cap-and-trade program's incentives for emission reductions, the reductions in emission intensity under H.R. 2454 partly reflect the effect of energy efficiency provisions in the bill.

Source: Energy Information Administration, *Energy Market and Economic Impacts of H.R. 2454, the American Clean Energy and Security Act of 2009*, August 2009.

Figure 7. EIA Projections of Changes in the CO_2 Emission Intensity of Production from 2006 to 2020 under Business-as-Usual Circumstances and under H.R. 2454 in Six Energy-Intensive Industrial Sectors.

Finally, it is important to recognize that impacts on production costs do not directly translate into impacts on industry competitiveness, profitability, and emission leakage. Even among those industries that experience the same percentage increase in production costs, differences in competitive conditions may allow some industries to pass more of those costs on to consumers than others, leading to varying impacts on industry profitability. Likewise, because profit margins (as a percent of revenue) can vary across industry, while the relatively small impacts on production costs depicted for most industries in Figure 5 may have little effect on some industries' profitability and ability to compete in a global marketplace, those same production cost impacts may more substantially affect the profitability and competitiveness of other industries where profit margins are lower.

B. Determinants of Leakage Resulting from Increased Domestic Production Costs

Many emission-intensive industries have various characteristics that make them relatively immobile in the face of small changes in production costs that might encourage firms to relocate to other countries.[19] Even among those emission-intensive industries that are considered trade-exposed, several factors can influence the extent to which a given industry is susceptible to competition from imports and to competition in export markets. As a result, in the face of similar increases in domestic costs, some trade-exposed industries may experience far smaller increases in imports or reductions in exports than others. A few examples of these factors are:

- Existing cost advantages: Due to access to inexpensive raw materials, advanced technologies, highly skilled labor, or other advantages, some domestic industries or firms may already enjoy a cost advantage relative to their international competitors that would remain even in the face of an increase in domestic production costs.[20]
- Fixed plant costs: Firms with a significant share of their investments in large, fixed physical structures, such as large manufacturing plants, may also be less sensitive to increases in production costs because the costs of relocation may outweigh the gains to locating in a less stringent regulatory environment.[21] For the same reason, they may be less exposed to international competition from other firms in the face of increased domestic costs if the capital investments required to build new capacity in foreign countries are large compared with the increase in domestic production costs.
- Transportation costs: Because of transportation costs, industries that produce products with a relatively low value per unit of weight, such as cement, would be less affected by a given increase in domestic production costs than would those that produce products with a relatively high value per unit of weight, such as steel. Likewise, because of transportation costs, even within an industry there may be geographic variation in the susceptibility of firms to international competition, with firms that serve markets near major ports being more susceptible to international competition than those that serve markets further inland.[22]
- Availability of spare international production capacity: In the near-term, an increase in domestic production costs may not have a significant effect on the competitive position of a domestic industry if globally there is little spare production capacity.[23]
- Uncertainty about future conditions: Many of the "presumptively eligible" sectors are capital intensive. As a result, foreign competitors may be reluctant to make significant investments in response to increases in production costs in the United States if there are significant uncertainties regarding future market conditions, including their domestic regulatory regimes.[24] Given the long lifetimes of these capital investments and the relative level of capital abundance in developed countries, a brief period of differentiation in domestic regulatory programs may not justify expansion by foreign competitors.[25]

- Product differentiation: In some industries, domestic producers may have an advantage of producing products that best meet domestic needs, either due to the unique needs of specific domestic customers, or due to regulatory product quality standards.[26]
- Agglomeration economies: While the sources of agglomeration economies are varied (e.g., knowledge spillovers, labor market pooling, proximity to firms that produce inputs or purchase outputs), their effect is the same — firms will have an incentive to locate near one another. Industries with significant agglomeration economies may be insensitive to increases in production costs if the gain from remaining close to other firms in the industry outweighs the gains from relocating to a region with less stringent greenhouse gas regulations.[27]

Any impacts of a cap-and-trade program on the international competitiveness of energy-intensive trade-exposed industries will occur against a backdrop of widely varying existing trade flows and trends therein. For some of the key industries that would likely be deemed eligible for allocations to "trade-vulnerable" industries under H.R. 2454, Figure 8 depicts trends in imports as a share of domestic consumption for the decade from 1998 to 2007, the most recent year for which domestic production data are available. Over this period, import penetration increased the most in the aluminum industry, which also had the highest level of import penetration throughout the period.[28] By contrast, the cement industry had the lowest import penetration throughout the period and, while that penetration rate varied throughout the period, in 2007 it was effectively the same as in 1998. Figure 9 depicts trends in exports of those industries.

If the United States adopts a cap-and-trade program, the extent to which this raises concerns about international competitiveness and leakage in a particular industry also depends critically on the specific countries that are sources of import or export competition. The key concern regarding a domestic cap-and-trade program's international competitiveness impacts is that it could put U.S. industry at a disadvantage relative to industry in those countries that do not adopt comparable regulations, such that manufacturing activity and emissions may shift to those countries. Yet, our trading partners in the European Union are already subject to mandatory greenhouse gas regulations, and several other major trading partners likely will adopt comparable regulations in the near future, particularly if the United States adopts a cap-and-trade program. For example, Canada and Mexico have discussed such coordination of cap-and-trade programs at the North American Leaders Summit this year. Thus, concern about leakage focuses only on trade with those countries that may not adopt comparable regulations in the near-term.

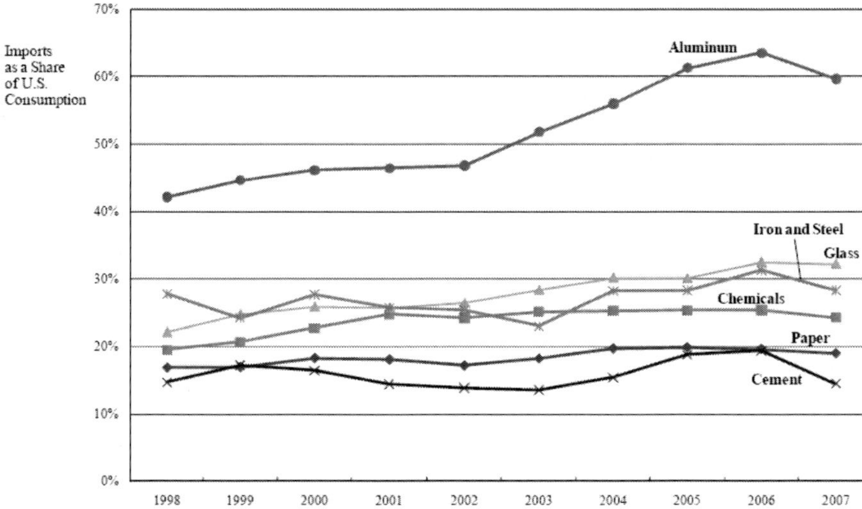

Note: Values are calculated as the dollar value of the sector's imports divided by the dollar value of the sum of its domestic production and net imports. The six-digit presumptively eligible sectors included in the sector categories are: Aluminum (Alumina Refining and Primary Aluminum Production); Cement (Cement Manufacturing); Chemicals (Petrochemical Manufacturing, Inorganic Dye and Pigment Manufacturing, Alkalies and Chlorine Manufacturing, Carbon Black Manufacturing, All Other Basic Inorganic Chemical Manufacturing, Cyclic Crude and Intermediate Manufacturing, All Other Basic Organic Chemical Manufacturing, Plastics Material and Resin Manufacturing, Synthetic Rubber Manufacturing, Cellulosic Organic Fiber Manufacturing, Noncellulosic Organic Fiber Manufacturing, and Nitrogenous Fertilizer Manufacturing); Glass (Flat Glass Manufacturing, Other Pressed and Blown Glass and Glassware Manufacturing, and Glass Container Manufacturing); Iron and Steel (Iron and Steel Mills); and Paper (Pulp Mills, Paper Mills, Newsprint Mills, and Paperboard Mills).

Source: U.S. Census Bureau and U.S. International Trade Commission data.

Figure 8. Imports as a Share of U.S. Consumption for Various "Presumptively Eligible" Industries.

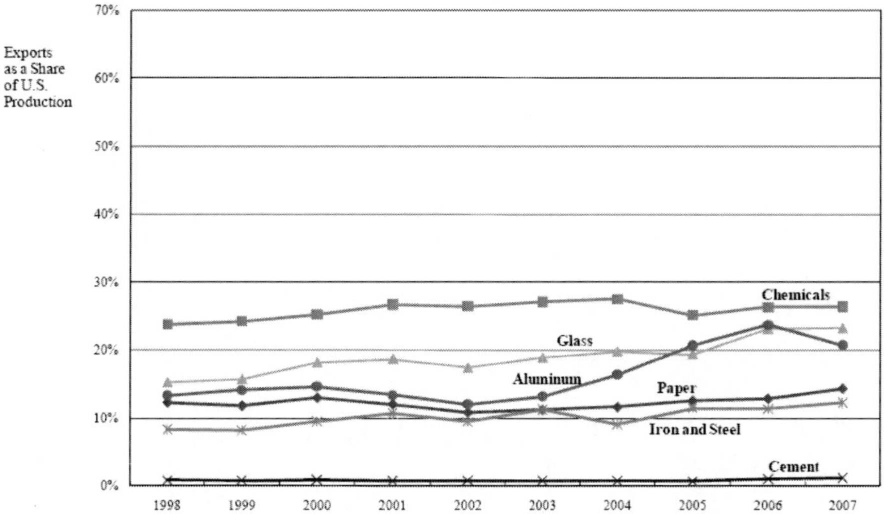

Note: Values are calculated as the dollar value of exports divided by the dollar value of the industry's domestic production. The six-digit presumptively eligible sectors included in the sector categories are: Aluminum (Alumina Refining and Primary Aluminum Production); Cement (Cement Manufacturing); Chemicals (Petrochemical Manufacturing, Inorganic Dye and Pigment Manufacturing, Alkalies and Chlorine Manufacturing, Carbon Black Manufacturing, All Other Basic Inorganic Chemical Manufacturing, Cyclic Crude and Intermediate Manufacturing, All Other Basic Organic Chemical Manufacturing, Plastics Material and Resin Manufacturing, Synthetic Rubber Manufacturing, Cellulosic Organic Fiber Manufacturing, Noncellulosic Organic Fiber Manufacturing, and Nitrogenous Fertilizer Manufacturing); Glass (Flat Glass Manufacturing, Other Pressed and Blown Glass and Glassware Manufacturing, and Glass Container Manufacturing); Iron and Steel (Iron and Steel Mills); and Paper (Pulp Mills, Paper Mills, Newsprint Mills, and Paperboard Mills).

Source: U.S. Census Bureau and U.S. International Trade Commission data.

Figure 9. Exports as a Share of U.S. Production for Various “Presumptively Eligible” Industries.

Figure 10 depicts the distribution of U.S. imports and exports by origin and destination for several “presumptively eligible” industries. Overall, half of our imports and half of our exports for these goods come from or go to the European Union, Canada, Australia, Japan, and New Zealand.

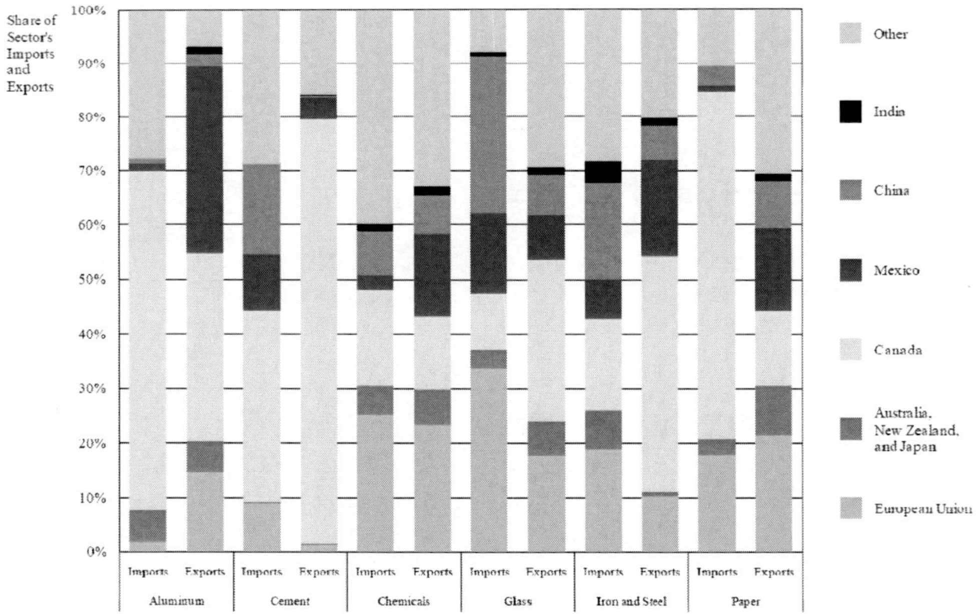

Note: Six-digit presumptively eligible sectors included in industry categories depicted above are: Aluminum (Alumina Refining and Primary Aluminum Production); Cement (Cement Manufacturing); Chemicals (Petrochemical Manufacturing, Inorganic Dye and Pigment Manufacturing, Alkalies and Chlorine Manufacturing, Carbon Black Manufacturing, All Other Basic Inorganic Chemical Manufacturing, Cyclic Crude and Intermediate Manufacturing, All Other Basic Organic Chemical Manufacturing, Plastics Material and Resin Manufacturing, Synthetic Rubber Manufacturing, Cellulosic Organic Fiber Manufacturing, Noncellulosic Organic Fiber Manufacturing, and Nitrogenous Fertilizer Manufacturing); Glass (Flat Glass Manufacturing, Other Pressed and Blown Glass and Glassware Manufacturing, and Glass Container Manufacturing); Iron and Steel (Iron and Steel Mills); and Paper (Pulp Mills, Paper Mills, Newsprint Mills, and Paperboard Mills).

Source: U.S. International Trade Commission data.

Figure 10. Distribution of U.S. Imports and Exports by Origin and Destination in 2008 for Various "Presumptively Eligible" Industries.

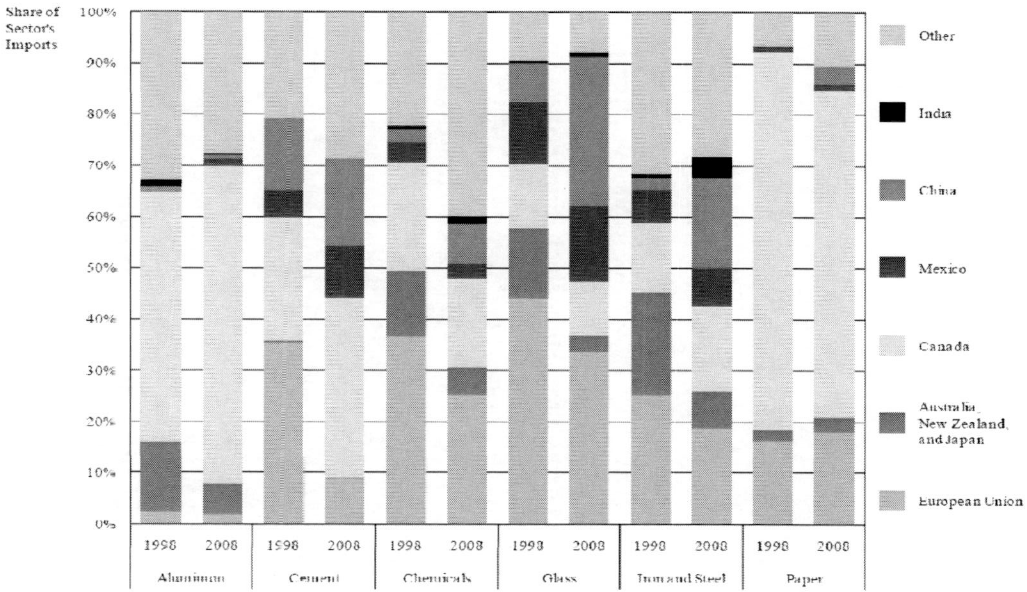

Note: Six-digit presumptively eligible sectors included in industry categories depicted above are: Aluminum (Alumina Refining and Primary Aluminum Production); Cement (Cement Manufacturing); Chemicals (Petrochemical Manufacturing, Inorganic Dye and Pigment Manufacturing, Alkalies and Chlorine Manufacturing, Carbon Black Manufacturing, All Other Basic Inorganic Chemical Manufacturing, Cyclic Crude and Intermediate Manufacturing, All Other Basic Organic Chemical Manufacturing, Plastics Material and Resin Manufacturing, Synthetic Rubber Manufacturing, Cellulosic Organic Fiber Manufacturing, Noncellulosic Organic Fiber Manufacturing, and Nitrogenous Fertilizer Manufacturing); Glass (Flat Glass Manufacturing, Other Pressed and Blown Glass and Glassware Manufacturing, and Glass Container Manufacturing); Iron and Steel (Iron and Steel Mills); and Paper (Pulp Mills, Paper Mills, Newsprint Mills, and Paperboard Mills).

Source: U.S. International Trade Commission data.

Figure 11. Distribution of U.S. Imports by Origin in 1998 and 2008 for Various “Presumptively Eligible” Industries.

However, as Figure 10 illustrates, there is significant variation in the pattern of trade across industries. In addition, there is substantial variation in trade patterns across specific industries within some of the broader sectors depicted in Figure 10, such as the chemicals sector. It should also be noted that, in some sectors, the share of trade that is with developing countries has increased over the past decade (see Figure 11), a trend that may continue into the future.

V. EXPERIENCE IN THE EUROPEAN UNION EMISSION TRADING SCHEME

The European Union Emission Trading Scheme (EU ETS) provides valuable lessons regarding a greenhouse gas cap-and-trade program's impacts on energy-intensive industry. The EU ETS covers the emissions from four broad sectors: iron and steel, certain mineral manufacturing industries (including the cement industry), energy production (including electric power facilities and direct emissions from oil refining), and pulp and paper. The first phase of the program ran from 2005 to 2007, and the program's emission target was tightened for a second phase from 2008 to 2012. The EU has also announced targets for a Phase III, which would mandate an emission reduction of at least 21 percent compared with 2005 levels by 2020.

Prospective studies of the EU ETS have shown that concern over the system's potential impacts on trade flows should be limited to a small set of industrial sectors. For example, a study of trade impacts in the U.K. identified six main sectors as either "significantly or plausibly of concern."[29] These include iron and steel, aluminum, nitrogen fertilizer, cement and lime, basic inorganic chemicals (e.g., chlorine and alkalies) and pulp and paper. EU-wide, a recent study found overall leakage of EU emissions is unlikely to be greater than 1 percent, but that impacts on EU production and imports could be significantly larger in the iron and steel and cement sectors.[30] At the same time, the literature on the EU ETS provides support for the conclusion that output-based allocations, like those proposed in H.R. 2454, could significantly mitigate the competitive impacts of a cap-and-trade program. For example, studying Europe's cement sector, Demailly and Quirion (2006) concluded that, if the EU cement producers received output-based allowance allocations at a rate equal to 90 percent of the industry's historical emission intensity, then imports to the EU would be "insignificantly impacted" under an EU cap-and-trade program even if allowance prices reached as high as €50 per ton of CO_2.[31] The study further found that the impacts on net imports become more significant (i.e., over 5 percent) when the output-based allocation rate is a lesser percentage of the industry's historical emission intensity (i.e., 75 percent or less). Finally, an analysis by the Carbon Trust in the U.K. found that, even in highly energy-intensive sectors such as blast furnace steel, output-based allocations can significantly reduce the increases in imports that could otherwise result from the EU ETS (Figure 12).[32]

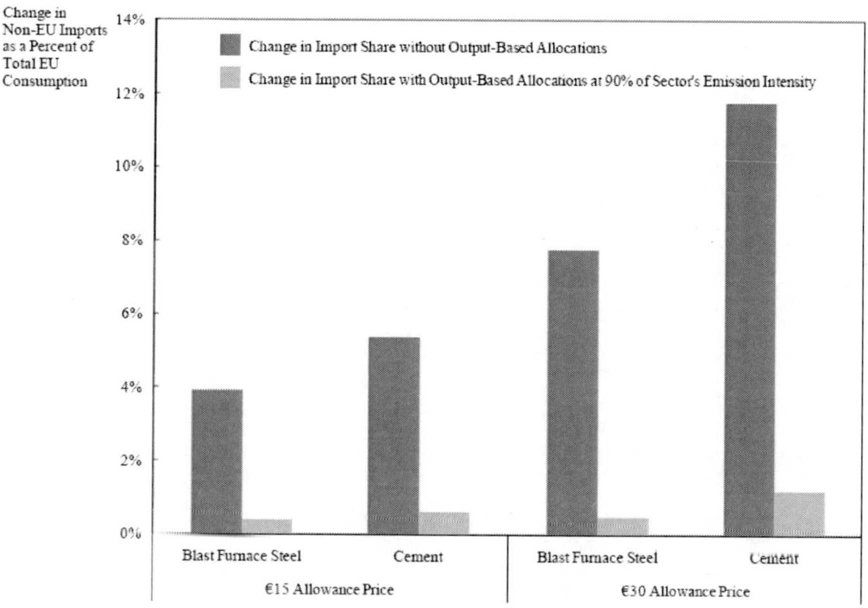

Note: Values presented here are derived from values for "complete pass-through" scenario presented in Charts 10 and 14 of the Carbon Trust report.

Source: Carbon Trust, *EU ETS Impact on Profitability and Trade: A Sector by Sector Analysis*, 2008.

Figure 12. Estimates of the Effect of Output-Based Allocations in Mitigating the European Union Emission Trading Scheme's Effect on Blast Furnace Steel and Cement Imports from Non-EU Countries under Alternative Allowance Prices.

While various studies have examined the potential for future impacts of the EU ETS on the international competitiveness of a limited set of European industries, it is important to note that retrospective studies of the first few years of the EU ETS have shown no competitive impacts from the program thus far. Baron et al. (forthcoming) examine data on imports to and exports from the EU in 2005 and 2006 in key energy-intensive trade-exposed sectors, including steel, cement, refining, and aluminum.[33] The authors find no statistical evidence that the EU ETS affected trade flows in those sectors. Moreover, the authors note that in sectors where imports increased (e.g., steel and aluminum), other factors, such as increases in consumption, may explain these increases. Other recent assessments of key sectors have found similar results.[34] Authors of these retrospective studies caution, however, that the early years of the EU ETS may not be representative of future impacts because of volatile and relatively low emission allowance prices, modest emissions targets, over-allocation of allowances to many facilities, long-term power contracts that precluded passing on all allowance costs, and other factors.

VI. Provisions in H.R. 2454 and Other Legislation that Can Mitigate Competitiveness Impacts

H.R. 2454 includes several provisions that could mitigate the potential impacts of a cap and-trade program on the international competitiveness of domestic manufacturing. These provisions are discussed below.

A. Allowance Allocations

1. Output-Based Allocations (or Rebates) to "Trade-Vulnerable" Industries

As was described earlier in this report, the most important measure included in H.R. 2454 to address competitiveness impacts in the near-term is the provision that would freely allocate (or rebate) emission allowances to certain energy-intensive sectors using a continuously updating output-based formula. Allowance rebates would be provided both for direct emissions — combustion emissions from fossil fuel use as well as process-related emissions — and for indirect emissions associated with the purchase of electricity. Rebates for direct emissions would be provided to the covered entities within an eligible sector (*i.e.*, those entities that are required to submit allowances for their direct emissions). Rebates for indirect emissions would be provided to each entity within an eligible sector regardless of whether or not that entity has an emission allowance compliance obligation under the cap-and-trade program.[35] In each year, rebates for direct emissions would be calculated by multiplying an eligible entity's average output in the two prior years by the average direct GHG emissions per unit of output for all entities within that entity's sector, where the latter estimate is updated every four years based on a four-year average. Rebates for indirect emissions would be calculated by multiplying an eligible entity's average output in the two prior years by the sector-average electricity use per unit of output (the "electricity efficiency factor"), which is updated every four years, and then multiplying the product by an entity-specific estimate of the emission intensity of that entity's electricity use in the prior year.[36] This last emission-intensity value is adjusted to account for any mitigation of rate impacts due to allocations to electricity local distribution companies (discussed further below). Each time they are updated with new data, the sector-wide direct GHG intensity value and the electricity efficiency factor described above are not permitted to increase relative to prior years, and the entity-specific value for the emission intensity of electricity consumption is not allowed to increase if that increase is the result of a change in the entity's source of electricity supply.

Allocations for indirect emissions associated with electricity purchases begin with the start of the cap-and-trade program in 2012. Allocations to eligible industrial sectors for their direct emissions begin in 2014, when the coverage of the cap-and-trade expands to begin including direct emissions from industrial facilities. The total number of allowances available for allowance rebates to eligible sectors for their direct and indirect emissions is equal to about 15 percent of the overall cap in 2014, and this number declines in tandem with the overall cap from 2014 to 2025. After 2025, H.R. 2454 specifies a default phase-out schedule for both the entity-specific allocations and the total pool of allowances set aside for these allocations. This default schedule would lead to a complete phase-out of those allocations by 2035, but that phaseout schedule can be delayed by the President based on considerations relating to the state of global action to reduce greenhouse gas emissions.

By freely distributing a specified number of allowances to firms for every additional unit of output they produce based on the average emission rate in the sector, these allocations offset the average effect that the cap-and-trade program would otherwise have on the production costs of recipients in each sector, and therefore on their international competitiveness and emission leakage. Importantly, the distribution of allowances within a sector on the basis of each firm's on-going production levels, rather than on the basis of their emissions or fuel use, maintains the incentives that a cap-and-trade program creates for each firm to reduce the emission-intensity of its production, and for market share within a sector to shift toward those firms that are the least emission intensive. Because all firms within a sector receive the same allocation per unit of their output and they still must surrender allowances to cover every ton of their emissions, even with these allocations, less emission-intensive firms still gain the same advantage over their more emission-intensive competitors as they would in a cap-and-trade program that does not provide such allocations. Output-based allocations simply shift downward the cap-and-trade program's effects on *all* firms' production costs on a consistent, industry-wide basis.

The primary benefit of output-based allocations is that they reduce emission leakage that can otherwise undermine the environmental effectiveness of a domestic cap-and-trade program, and at the same time they avoid the economic losses associated with that leakage. Yet, because these allocations accomplish this by mitigating the cap-and-trade program's effects on domestic production costs of emission-intensive goods, these allocations also diminish incentives for consumers to reduce their use of emission-intensive goods, and thereby forego some cost-effective domestic emission reductions. The allocations also involve the free distribution of allowances whose value could instead be devoted to other uses, including support for low-income consumers, clean energy technology, or deficit reduction. To minimize overall costs, it is therefore important to narrowly target these allocations to trade-exposed goods and to choose the level of allocation that balances these costs and benefits.

2. Allocations to Electricity and Natural Gas Local Distribution Companies (LDCs)

Provisions to distribute allowances to electricity and natural gas local distribution companies (LDCs) would also benefit energy-intensive trade-exposed industries. H.R. 2454 allocates about 30 percent of the value of allowances to electricity LDCs annually from 2012 to 2025, distributing them ratably among LDCs, 50 percent on the basis of emissions, and 50 percent on the basis of electricity deliveries, with the delivery component allocated on an updated basis every 3 years from 2015. The additional 9 percent of allowances given annually to natural gas LDCs from 2016 to 2025 are allocated ratably on the basis of historical deliveries, with an adjustment every 3 years starting in 2019 to account for changes in the size of each utility's customer base. Emission allowances allocated to electricity and natural gas LDCs must be used for the benefit of retail ratepayers, and ratepayer benefits must be distributed ratably among ratepayer classes based on deliveries to each class.

In providing rebates to residential and commercial customers, LDCs are instructed not to provide rebates that are based solely on the quantity of electricity or natural gas delivered, which would amount to rate reductions. Among other potential approaches, one alternative to such a rebate would be to provide a fixed rebate that is unrelated to a customer's electricity consumption. LDCs are, however, allowed to use allocations to reduce industrial customers' electricity rates.

As with the output-based rebates, the LDC allocations offset a portion of the effect that the cap-and-trade program would otherwise have on the production costs of domestic manufacturers, and therefore mitigate competitiveness impacts. Indirect emissions from electricity consumption account for more than one-quarter of the total emissions of the "presumptively eligible" energy-intensive trade-exposed industries. Moreover, EPA analysis suggests that the size of the allocations to electricity LDCs during the coming decade would be on the order of 80 percent of the electricity sector's emissions. Therefore, the allocation to electricity LDCs could mitigate a significant portion of the costs that energy-intensive trade-exposed industries might otherwise experience under a cap-and-trade program. Further, because the output-based allocations to eligible energy-intensive trade-exposed industries would be provided for both direct emissions *and* indirect emissions associated with the purchase of electricity, any residual impacts on electricity costs left unaddressed by the LDC allocation should be offset by the output-based allocations.[37] On the other hand, the allocation to natural gas LDCs would likely provide minimal benefit to the energy-intensive trade-exposed industries, as this allocation will only benefit natural gas users that receive their gas from LDCs and whose emissions are not directly regulated under H.R. 2454. Most energy-intensive trade-exposed manufacturers likely would be covered entities whose direct emissions will be regulated. Thus, such manufacturers would be ineligible to benefit from the natural gas LDC allocation, although their use of natural gas resulting in direct emissions would still enable them to receive the output-based allocations.[38]

Some have raised concerns that energy-intensive industry and other electricity consumers will not benefit from the allocations to LDCs. In particular, some have pointed to the experience in the EU ETS, in which electricity rates rose despite significant allowance allocations to electric generators, leading many generators to realize windfall profits from those allocations.[39] However, the proposed LDC allocations fundamentally differ from the EU ETS allocations. In particular, the LDC allocations would go to electric utilities that, by law, are subject to economic regulation by public utility commissions that dictate the rates those utilities can charge and the profits they can earn. Thus, state (or local) public utility commissions will have the ability to dictate how the value of allocated emission allowances is used to benefit ratepayers. By contrast, the EU ETS freely allocated allowances to electricity generators, many of whom operate in competitive electricity markets and are not subject to economic regulation by utility commissions. Because these allocations were fixed at the outset of the EU ETS, the resulting increase in electricity rates and windfall profits in the EU ETS were consistent with what economic theory would predict.[40]

Although the LDC allocations are specifically structured and targeted to ensure that ratepayers — rather than electric utilities — ultimately benefit from them, it is difficult to predict exactly how the value of those allocations will be distributed across ratepayers. These distributional decisions will be made by each LDC and the regulatory commission that oversees it. As a result, similar firms in the same energy-intensive trade-exposed industry that are served by different LDCs may experience very different benefits from the LDC allocations. In addition, while the LDC allocations can help achieve particular distributional objectives, these distributional benefits come at the expense of increasing the nationwide cost of the cap-and-trade program.[41] Specifically, to the extent that LDCs use the allocations in ways that reduce electricity rates for their customers, or that are perceived to do so, this will diminish the incentives that consumers face to reduce their electricity use or to rely on self-generated electricity, such as through combined heat and power. By diminishing these incentives, the allocations will reduce the cap-and-trade program's ability to realize some cost-effective emission reductions associated with electricity conservation and onsite electricity generation, and will thereby require greater reliance on other, more costly emission reductions to achieve the national emissions cap. Thus, in evaluating and designing LDC allocations, it is important to consider the complexities involved in achieving desired distributional objectives through LDC allocations, as well as the tradeoffs between the distributional benefits and economic costs of those allocations.

3. Comparison of Allocations with Emissions of "Presumptively Eligible" Industries

In establishing the output-based allowance rebates to energy-intensive trade-exposed industries, as noted above, H.R. 2454 limits the overall number of allowances that are set aside for these allocations. While the bill uses an allocation formula for eligible firms that is based on the average emission-intensity of their industry's production, if more allowances are needed in sum to fulfill this allocation formula across all eligible industries than are set aside for the allocations, each firm's allocation would be pro-rated down accordingly. If this were to occur, firms would receive allocations at a rate below the average emission-intensity of their industry's production.

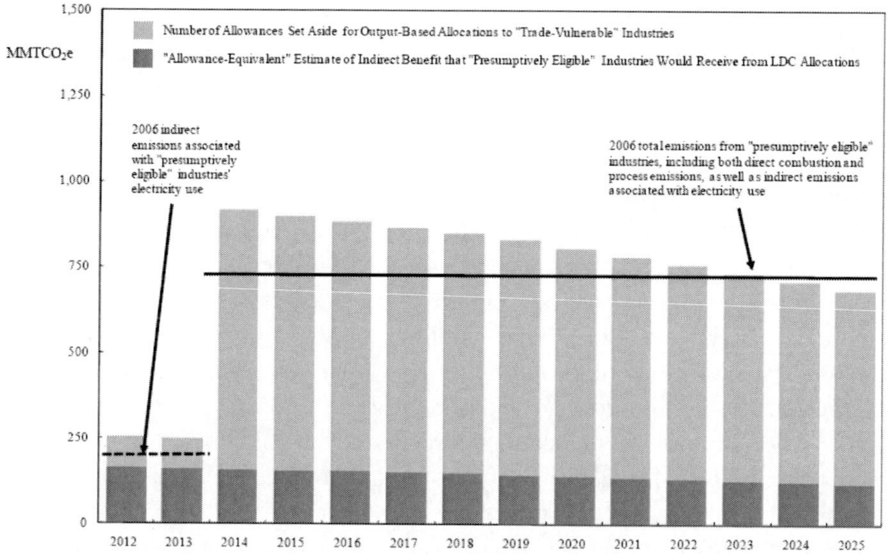

Note: Emission estimates are from Table 1. H.R. 2454 does not begin to cover these industries' direct emissions until 2014. The horizontal line in 2012 and 2013 therefore reflects their indirect electricity emissions only. While the maximum number of allowances available for output-based allocations to "trade-vulnerable" industries is known with certainty, the indirect benefit that these industries will receive from the LDC allocations is uncertain. The estimated "allowance-equivalent" value of the indirect benefit from the LDC allocation shown here is calculated by multiplying the total LDC allocation by the "presumptively eligible" industries' share of total 2007 electrictity sales in the United States, based on the 2007 Economic Census and EIA data on nationwide retail electricity sales

Figure 13. Comparison of the "Presumptively Eligible" Industries' 2006 Emissions with the Number of Allowances Set Aside for Output-Based Allocations and the Estimated "Allowance-Equivalent" of the Indirect Benefits from Allocations to Electricity Local Distribution Companies in H.R. 2454.

Figure 13 compares the total direct and indirect emissions of the "presumptively eligible" industries in 2006 with the combination of the total number of allowances that H.R. 2454 sets aside for output-based allocations to those industries, and an estimate of the "allowance-equivalent" benefit that those industries would receive from the electricity LDC allocations. Figure 13 presents this comparison through 2025. After 2025, the number of allowances that are set aside for the output-based allocations will depend in large part on Presidential determinations regarding the timing of the phase-out of those allocations.

Figure 13 demonstrates that, taking into account the potential to carry over unused allowances from one year to the next, as H.R. 2454 provides for, the number of allowances available through the LDC allocations and output-based allocations would exceed the 2006 emissions of the eligible industries through 2025. Of course, the adequacy of the allowance set asides will ultimately depend on how those set asides compare with the *future* emissions of the eligible industries. As was noted in Section III.B, anticipated continued reductions in the emission-intensity of the eligible industries' production methods will tend to reduce the future emissions of those industries relative to recent levels. On the other hand, findings that additional sectors would be eligible for allocations (e.g., through "individual showings") and increases in eligible industries' production levels relative to 2006 levels (i.e., relative to levels prior to the recent economic recession) would increase their future emissions. While projections of the total greenhouse emissions of the "presumptively eligible" industries are unavailable, as noted previously, EIA projects that the total energy-related CO_2 emissions of the six key energy-intensive trade-exposed sectors depicted in Figure 7 would decline nearly 20 percent from 2006 to 2020 under business-as-usual circumstances, and nearly 30 percent under H.R. 2454.

B. International Reserve Allowance Program

H.R. 2454 also includes an international reserve allowance program provision that could be applied to imports of emission-intensive goods from foreign producers that have not been subject to comparable greenhouse gas regulations. In particular, importers of energy- and emission-intensive goods may be required to surrender "international reserve allowances" to reflect the greenhouse gas emissions associated with the production of those goods, adjusted downward as appropriate to reflect any benefit that domestic producers are receiving from the output-based allocations provided to "trade-vulnerable" industries and from the LDC allocations. Under H.R. 2454, beginning in 2020, if a multilateral agreement on climate change has not been reached, this international reserve allowance program could go into effect for covered industrial sectors where 15 percent or more of imports are from countries that do not meet particular criteria relating to their greenhouse gas regulations, emission-intensity, and emission levels.

While both output-based allocations and the international reserve allowance program can minimize the differential in product prices between energy-intensive goods produced in countries with greenhouse gas regulations and those produced elsewhere, the two policy mechanisms do so in very different ways. Output-based allocations would be likely to lower the cost of domestic production in a country with greenhouse gas regulations, offsetting the price increase for domestically produced goods that would otherwise occur under a cap-and-trade regime. This should minimize differences in product prices for domestically-produced and foreign-produced goods in both domestic and international markets.

The international reserve allowance program in H.R. 2454, on the other hand, would be likely to raise the cost of certain imports from countries without comparable greenhouse gas regulations, allowing both domestic and imported covered goods to carry a carbon price that is intended to reflect the emissions associated with their production, raising the price of both domestic and foreign energy-intensive goods in domestic markets. It is also important to consider that an international reserve allowance provision could raise possible issues of consistency with international obligations that could result in retaliatory actions by trading partners that could result in higher costs for the export of U.S. goods.

C. Cost Containment Mechanisms

Cost containment mechanisms can also mitigate competitiveness impacts on energy-intensive trade-exposed industries by reducing the overall cost of complying with a cap-and trade program.

H.R. 2454 and other recent climate proposals include several provisions that can lower the overall costs of a cap-and-trade program, including banking and borrowing of allowances and emissions offsets. In the case of offsets, EPA analysis of H.R. 2454 has shown that the use of domestic and international offsets can reduce the cap-and-trade program's overall costs, assuming appropriate institutional capacity.[12]

D. Additional Provisions That Benefit Manufacturing

Several additional provisions in H.R. 2454 would either directly or indirectly improve the international competitiveness of energy-intensive trade-exposed industries:

- *Clean Energy Deployment Administration.* The bill would authorize \$7.5 billion in bonds to initially capitalize a Clean Energy Deployment Administration (CEDA) — commonly referred to as the "Green Bank" — to promote domestic deployment of clean energy technologies. Among its several goals would be the promotion of "domestic production of commodities and materials (such as steel, chemicals, polymers, and cement) using clean energy technologies so that the United States will become a world leader in environmentally sustainable production of the commodities and materials." The CEDA would be authorized to pursue its goals through providing both direct and indirect financial support. Direct support would include direct loans, letters of credit, and loan guarantees, while indirect support could include credit support and facilitation of certain financing transactions.
- *Bonus allowances for carbon capture and sequestration (CCS) at industrial facilities.* Starting in 2014, H.R. 2454 sets aside allowances — reaching 5 percent annually after 2020 — to distribute to entities that deploy commercial-scale technologies that capture and store at least 50 percent of potential carbon emissions. Up to 15 percent of these CCS bonus allowances can be awarded to industrial sources. Such bonus allowances could reduce the costs that energy-intensive industries face to reduce their emission-intensity.
- *Clean Energy Manufacturing Revolving Loan Fund.* This provision, based on Senator Sherrod Brown's Investments for Manufacturing Progress and Clean Technology (IMPACT) proposal, would provide incentives to small- and medium-sized manufacturers for the production of clean energy technologies and for the improvement of energy efficiency. The Department of Energy would be given \$15 billion a year in 2010 and 2011 to award grants to States to establish revolving loan funds that would provide loans to small- and medium-sized manufacturers. These loans could finance the cost of reducing the energy intensity or greenhouse gas emissions of an eligible facility. Loans could also finance the cost of reequipping, expanding,

or establishing a manufacturing facility to produce clean energy technology or energy-efficient products.

- *Clean energy and efficiency manufacturing partnerships.* H.R. 2454 would authorize an average of $300 million to be appropriated annually for the next 5 years to the Secretary of Commerce for the Hollings Manufacturing Partnership Program and would reduce the cost share requirement for participating manufacturers. This partnership program provides support to manufacturers to undertake clean energy manufacturing initiatives, including "reducing energy intensity and greenhouse gas production, including the use of energy intensive feedstocks."
- *Worker adjustment assistance and training.* H.R. 2454 would distribute 0.5 percent of the value of allowances annually to finance worker adjustment assistance. Workers "employed in ... energy-intensive manufacturing industries" are eligible to apply for such assistance, which would include adjustment benefits and new job training.

E. Provisions from the American Recovery and Reinvestment Act of 2009 That Benefit Manufacturing

In addition to the provisions in H.R. 2454, several provisions in the American Recovery and Reinvestment Act (ARRA) of 2009 benefit energy-intensive trade-exposed sectors by supporting retooling and increased efficiency at industrial facilities and by creating demand for energy intensive primary materials (*e.g.*, steel for windmill blades). Relevant investments and incentives provided by ARRA include:

- *Advanced Energy Manufacturing Facility Investment Tax Credit.* ARRA authorizes up to $2.3 billion in tax credits for qualified investments to re-equip, expand, or establish facilities designed to manufacture advanced energy technologies including, among others: renewable energy technologies, fuel cells, microturbines, energy storage systems for electric/hybrid vehicles, energy efficiency technologies, smart grid equipment, plug-in hybrid vehicles, and equipment to capture and sequester carbon dioxide. The tax credit would be equal to 30 percent of the value of the qualifying investment.

- *Clean Renewable Energy Bonds.* ARRA authorizes an additional $1.6 billion of bonds to finance facilities that generate electricity from wind, closed-loop biomass, open-loop biomass, geothermal, small irrigation, hydropower, landfill gas, marine renewable, and municipal waste combustion facilities.
- *Advanced Battery Manufacturing Grants.* ARRA allocates $2.0 billion for grants to U.S. manufacturers of advanced battery systems and car batteries. Funding could go toward the production of lithium ion batteries, hybrid electrical systems, and system components.
- *Loan Guarantees for Renewable Energy and Transmission.* ARRA allocates $4.0 billion for a program to deploy renewable energy and electric power transmission. This amount is expected to leverage as much as $40 billion in loan guarantees for investments in renewable energy projects and related transmission systems.
- *Industrial Energy Efficiency.* The Department of Energy has announced the award of $156 million in ARRA funds for energy efficiency improvements in major industrial sectors. The program is designed to reduce energy consumption in manufacturing and information technology sectors. One of the major focuses of the program will be investment in technologies that will reduce energy costs and make U.S. industry more competitive. Among the projects to be funded will be combined heat and power projects, district energy systems, waste energy recovery systems, and efficient industrial equipment.
- *Payments for Specified Energy Property in Lieu of Tax Credits.* ARRA authorizes the Department of the Treasury to make payments to eligible persons that either place in service particular types of energy property in 2009 or 2010, or, under certain circumstances, begin construction on such property in 2009 or 2010. These payments are generally equal to 10 or 30 percent of the eligible cost basis of the property, depending on the type of property. Those that apply for these payments must forego tax credits that they could otherwise receive under sections 45 and 48 of the Internal Revenue Code. The types of energy property that are eligible for these payments include, among others, various renewable energy generation facilities, microturbines, and fuel cells.
- *Training.* ARRA includes several provisions that would provide funding for job training, including $500 million for research, labor

exchange, and job training projects that prepare workers for careers in energy efficiency and renewable energy.

VII. Economic Modeling of Competitiveness Impacts and Associated Leakage

While the United States does not yet have experience with a greenhouse gas cap-and trade program to draw on for retrospective studies of emission leakage and associated competitiveness impacts, the economics literature has employed several analytical approaches to assess the anticipated effects of climate policy on domestic energy-intensive trade-exposed manufacturing industries. These studies generally find that, on average, impacts are likely to be modest. This section briefly reviews prior economic analyses of these issues and then describes and presents the results from new modeling performed for this report.

Prior analyses by Aldy and Pizer, and by Ho, Morgenstern, and Shih offer examples of two different approaches that have been used to examine competitiveness impacts, the former relying on a statistical approach that uses historical data, and the latter using simulation modeling.[42] Notably, both studies focus on competitiveness impacts that would exist *in the absence of* any mitigating policy measures like those incorporated in H.R. 2454. Aldy and Pizer examine the past impacts of higher domestic energy prices on energy-intensive trade-exposed industries in order to predict how those industries might be affected by a $15 per ton of CO_2 emissions allowance price in 2012. They estimate that the short-run decline in domestic production in the iron and steel, aluminum, cement, glass, and paper industries will be relatively small, ranging between 1.6 and 3.4 percent, of which only a portion would reflect the impacts of reduced international competitiveness, as is described further below.[43] Ho, Morgenstern, and Shih also examine the short-run impacts of a cap-and-trade program on these industries, focusing on a $10 per ton of CO_2 allowance price. They find that most energy-intensive trade-exposed industries would experience a drop in domestic output of between 1 and 3 percent, but that a limited set of those industries would experience greater impacts. For example, they project that cement manufacturers would experience a 4.1 percent decline in domestic production and that petrochemical manufacturers would experience a 7.7 decline in domestic production.[44] The authors estimate that short-run impacts on industry profits will be even smaller, falling by less than one percent in each of the energy-intensive trade-exposed industries that they examine. The authors further project smaller impacts on domestic production in the long run, when the economy can fully adjust to climate policy by investing in low-carbon equipment and processes. While these studies use different methods, they reach common conclusions that, on average, the effects of a cap-and-trade program on domestic production levels of energy-intensive trade-exposed industries are modest.

These studies also highlight that the changes in the domestic production levels of energy-intensive trade-exposed industries under a cap-and-trade program will not solely arise from changes in international competitiveness. Rather, some — and, for certain industries, most — of the change in their domestic production levels will result from a decrease in the domestic consumption of their products as the U.S. economy moves to less emission-intensive goods, not from a loss of international competitiveness. For instance, Aldy and Pizer find that, for some industries, absent any mitigating policy measures, changes in international competitiveness could account for as little as one-fifth of the change in domestic production levels under a cap-and trade program, but that it could account for a greater share in other industries. Ho, Morgenstern, and Shih find that changes in international competitiveness would account for between 40 and 50 percent of the decrease in domestic production of primary metals, nonmetallic minerals, and chemicals and plastics.

It is also notable that, although impacts will be more significant in some industries than in others, on the whole across all industries these studies project small declines in U.S. manufacturing output and very low levels of associated leakage. Aldy and Pizer find that a $15 allowance price would have virtually no impact on manufacturing employment and would lead to a 1.3 percent decline in total manufacturing output, with about half of this decline in output associated with a shift in production abroad.

To examine a domestic cap-and-trade program's effects on the international competitiveness of, and emission leakage from, energy-intensive trade-exposed industries, and to assess the efficacy of provisions in H.R. 2454 to address those impacts, we have drawn on both existing modeling by EPA and EIA, as well as new modeling conducted for this report. Specifically, based on the relative strengths of the respective models, the EPA and EIA modeling were used as a basis for assessing the broad economy-wide effects of a cap-and-trade program, including the allowance prices that would be expected to emerge from a cap-and-trade program like that in H.R. 2454. Using this foundation, a multi-region, multi-sector model of the global economy was employed to examine international competitiveness impacts of a domestic cap and-trade program on several energy-intensive trade-exposed industries under various policy scenarios. These policy scenarios differed according to the degree of global action to address climate change, and according to the domestic measures that are adopted to mitigate competitiveness impacts.

The new modeling in this report was performed using the Fischer-Fox Emissions and Trade (FFEAT) model, which is based on the GTAP 7 database[45] of the global economy. While producing results that are consistent with prior EPA and EIA modeling, the new modeling conducted for this report builds on that prior modeling in ways that make it possible to glean various insights that cannot be obtained from the EPA or EIA modeling alone. For example, the new modeling makes it possible to examine climate policy and trade impacts on industrial sectors at a more disaggregated level than is possible in the EPA modeling. In addition, while the EIA modeling can also examine impacts on industrial production at a more disaggregated level than is possible in the EPA modeling, unlike the EIA modeling, the modeling conducted for this report makes it possible to differentiate those impacts on specific sectors' production levels that are attributable to changes in international competitiveness from those that are attributable to changes in domestic demand for emission-intensive products. Moreover, the new modeling conducted for this report examines certain policy scenarios that were not examined in prior EPA and EIA modeling, allowing for an improved understanding of how competitiveness impacts are influenced by different levels of global action to address climate change, and by particular domestic policy measures.

The remainder of this section provides an introduction to the model that was employed, describes the scenarios that were examined, details the particular industrial sectors that were evaluated, and then presents the conclusions from the modeling exercise.

A. Overview of Economic Model

The FFEAT model is a refinement of a model that has been used in prior peer-reviewed research.[46] It is similar to EPA's ADAGE model, which EPA employed in analyzing H.R. 2454, in the sense it accounts for both the direct and indirect (or "general equilibrium") effects of a cap-and-trade program on both the domestic economy and international trade flows.[47] However, the

FFEAT model differs from ADAGE in several important respects. On the one hand, it provides greater sectoral disaggregation, which enhances efforts to examine impacts on specific energy-intensive trade-exposed industries. On the other hand, it does not have the capacity to examine the effects of a cap-and-trade program that develop over time, such as the cumulative effect of investments in response to a policy change. Rather, it focuses on initial effects of a policy change. Additionally, the FFEAT model does not account for many features of a cap-and-trade program that are necessary to reliably project the allowance price that would prevail in a given year under a program like that in H.R. 2454.[48] Nonetheless, if a particular allowance price is assumed based on other modeling efforts, such as those by EPA and EIA, the FFEAT model can be used to examine the economic implications of that allowance price on energy-intensive trade-exposed industries.

The FFEAT model accounts for some of the potential adjustments to the production processes of industry — and economic activity more broadly — that would be expected in response to a cap-and-trade program. The kind of adjustments that are accounted for are those that can be expected over a "medium-run" time horizon. From the perspective of energy-intensive trade-exposed industries, if climate legislation were enacted in the coming months, this time horizon is broadly consistent with the time that would elapse between the enactment of legislation and the initial years in which those industries' emissions would be covered under H.R. 2454's cap-and-trade program.

Unlike other models that evaluate a policy's effects against a continually adjusting "baseline" projection of how the economy would otherwise evolve into the future, the FFEAT model evaluates a policy's effects as if it were imposed on the structure of the global and domestic economy that existed in 2004. To the extent that the economy evolves in important ways between 2004 and the year in which a domestic climate policy is implemented, the FFEAT model can over- or understate the relative impacts on particular sectors or countries. For example, the model will tend to overstate impacts on those sectors that experience a significant decline in emission intensity (*e.g.*, a decline arising from technological progress) between 2004 and the implementation of a domestic cap-and-trade program, especially to the extent that the emission intensities of these sectors fall relative to their foreign competitors. Moreover, while the FFEAT model accounts for adjustments in the use of existing technologies and other production inputs in response to a climate policy, it does not account for potential technological change that could be spurred by a domestic climate policy. By using a 2004 baseline, the FFEAT model also excludes the effects that recent legislation — including the

Energy Independence and Security Act of 2007 and the American Recovery and Reinvestment Act of 2009 — would have in increasing the efficiency of manufacturing and growing a clean energy economy. On the other hand, due to the aggregated nature of the sectors modeled in the GTAP framework, the model may understate impacts on more narrowly-defined industries that are particularly emission intensive. Another important caveat is the increasingly important role of international trade in the U.S. economy, particularly trade with developing countries. As trade intensity increases over time, increasingly trade-exposed sectors are likely to see greater impacts from a policy change than the modeling presented here would suggest, especially in scenarios where other countries do not adopt comparable climate policies.

While these various factors can influence the magnitude of certain policy impacts that are identified in this modeling, they are unlikely to substantially influence the modeling's central finding: that measures like those incorporated in H.R. 2454 can effectively mitigate a cap-and trade program's impacts on domestic energy-intensive trade-exposed industries even if those industries face competition from firms in countries that have not yet adopted comparable regulations.

B. Policy Scenarios

As noted above, unlike models used by EPA and EIA, the FFEAT model currently cannot estimate the allowance price that would emerge under a cap-and-trade program that, like the one proposed in H.R. 2454, provides for the use of emission reduction offset credits and allowance banking. As a result, the analysis presented here simply assumes an allowance price of $20 per ton in order to examine a cap-and-trade program's impacts on domestic manufacturing. The $20 allowance price has been adopted for illustrative purposes in examining competitiveness impacts, and does not reflect a specific projection of allowance prices under H.R. 2454. However, it is broadly consistent with EPA, Congressional Budget Office, and EIA projections of allowance prices under H.R. 2454 in the latter half of the next decade, once energy-intensive trade-exposed industries would be fully covered by the cap-and-trade program. The FFEAT model is then used to explore the implications of different policy scenarios that are defined by the extent of global adoption of greenhouse gas regulations and by the measures taken at home to address competitiveness concerns.

1. Levels of Global Action

To examine a domestic cap-and-trade program's impact on the international competitiveness of U.S. energy-intensive trade-exposed industries, we examine a few possible scenarios of international engagement in reducing greenhouse gas emissions.

- *United States and Developed Country Action*: This scenario — the primary one on which we focus — offers a realistic representation of the near-term outlook for global engagement in addressing climate change. In line with past EPA analyses, in this scenario it is assumed that, with the exception of Russia, all other countries that were assigned emission targets under the Kyoto Protocol (Annex I countries) adopt mandatory emission targets. Indeed, as participants in the European Union Emission Trading Scheme, the majority of these countries have already adopted mandatory limits. It is further assumed that the emission allowance price in the United States and Annex I countries is the same, at $20 per ton. This common allowance price could emerge if each country develops a cap-and-trade program and those programs are linked either directly or indirectly, such as through the use of the same pool of international offset credits.
- *Unilateral United States Action*: This scenario is presented to provide a basis for comparison in evaluating how the international competitiveness impacts of a domestic climate policy on U.S. industry are mitigated by the adoption of greenhouse gas regulations in other developed countries, such as the European Union.
- *Developed Country Action without the United States*: Like the prior scenario, this scenario is used only to better understand international competitiveness impacts. Specifically, it is used to isolate the effects of the United States adopting a cap-and-trade program in the context of the primary policy scenario in which both the United States and other developed countries act.

2. Domestic Measures to Mitigate Competitiveness Concerns

The modeling presented here specifically examines the effect of two provisions in H.R. 2454 that can reduce a cap-and-trade program's impacts on the international competitiveness of energy-intensive trade-exposed industries: allocations to electricity LDCs and the output-based allocations (or rebates) to eligible "trade-vulnerable" industries. While other provisions in H.R. 2454 could also influence the international competitiveness of domestic industry, these two provisions are the most economically significant and could be incorporated in the current modeling framework in the available timeframe for this exercise.

In modeling the electricity LDC allocations in this exercise, it is assumed that the level of those allocations is equal to about 80 percent of the electricity sector's emissions. This is generally consistent with EPA's findings regarding the relative level of those allocations during the first decade of H.R. 2454's operation — the period of interest in this exercise. Furthermore, it is assumed that the LDC allocations would have the effect of reducing the electricity rates that industrial consumers face, as is allowed for in H.R. 2454.

In modeling the output-based allocations to eligible industrial sectors, consistent with H.R. 2454, we assume that firms in each eligible sector would receive a per-unit-of-output allocation that is equal to their sector's average emission intensity of production, considering both their sector's direct emissions as well as indirect emissions from electricity consumption. That is, we assume that a sufficient number of allowances are set aside to fulfill the allocation formulas for "trade-vulnerable" industries established in H.R. 2454.[49] Further, as provided for in H.R. 2454, in this modeling, the output-based allocation for indirect emissions are reduced to reflect the benefit that firms receive through the LDC allocations, so that the indirect emissions of the eligible industries are fully covered by a mix of LDC and "trade-vulnerable" industry allocations.

C. Energy-Intensive Trade-Exposed Industries Examined in Modeling

The GTAP modeling framework employed by FFEAT allows for the examination of a policy's impact on 50 aggregated economic sectors. Of these, in the context of this modeling, the five sectors considered energy intensive trade-exposed industries that would be eligible for output-based allocations are: chemicals, rubber, and plastics; pulp, paper, and print; nonmetallic minerals; iron and steel; and nonferrous metals. These aggregated sectors encompass 37 of the six-digit NAICS industries that are likely to be deemed "presumptively eligible" for output-based allocations, as described in Section III (see Table 2). In turn, these 37 industries account for 92 percent of the emissions and output of the "presumptively eligible" industries. The remaining nine industries fall within aggregated GTAP sectors that are not considered to be energy-intensive trade-exposed industries for purposes of the modeling exercise because those nine industries represent such a small fraction of the aggregated GTAP sectors in which they fall.

Because of the aggregation of economic activity in the GTAP database, the sectors considered to be energy-intensive and trade-exposed in the modeling include additional industries beyond just the narrowly-defined "presumptively eligible" six-digit NAICS industries that they include. In certain GTAP sectors, such as iron and steel, the "presumptively eligible" industries account for nearly two-thirds of the economic activity represented by that GTAP sector. However, in other sectors — such as paper, pulp, and print, and nonferrous metals — the "presumptively eligible" industries account for less one-quarter of the economic activity represented by the GTAP sector.

As a result of the aggregated nature of the GTAP sectors, their GHG and trade intensities tend to be lower than the corresponding intensities of the "presumptively eligible" industries within them. Consequently, this particular feature of the modeling may understate the average impacts on the international competitiveness of the "presumptively eligible" industries that would be expected to result from a cap-and-trade program without any mitigating policy measures. As noted previously, other features of the modeling can lead it to overstate these impacts.

D. Assessment of a Cap-and-Trade Program's Impacts on International Competitiveness and Associated Leakage under Alternative Policy Scenarios

A cap-and-trade program's impacts on the international competitiveness of energy-intensive trade-exposed industries and the resulting emission leakage flow from the program's effects on the production costs of those industries. The greater a cap-and-trade program's net impact on domestic industries' production costs relative to their foreign competitors, the greater will be the resulting competitiveness impacts and leakage. In light of this, the discussion below first focuses on the net impacts of a cap-and-trade program on the production costs of domestic industries, as well as the efficacy of both the output-based allocations and the LDC allocations in mitigating those impacts. The discussion then proceeds to examine the changes in trade flows and emission leakage that would result without the output-based allocations and LDC allocations, and then those changes that would result with those allocations.

Table 2. GTAP Energy-Intensive Trade-Exposed Sectors and Corresponding "Presumptively Eligible" Six-Digit NAICS Industries

GTAP Energy-Intensive Trade-Exposed Sector	GTAP Sector's Trade Intensity[1]	GTAP Sector's GHG Intensity[1]	Estimate of "Presumptively Eligible" Industries' Share of Corresponding GTAP Sector's Output	"Presumptively Eligible" Industries in Corresponding GTAP Sector[2]	Industry's Trade Intensity[1]	Industry's GHG Intensity[1]
Chemicals, Rubber and Plastics	33.W.	1.3%	29.0%	325110: Petrochemical Manufacturing	16.8%	1.6%
				325131: Inorganic Dye and Pigment Manufacturing	43.4%	1.5%
				325181: Alkalies and Chlorine Manufacturing (incl soda ash beneficiation)	29.2%	5.0%
				325182: Carbon Black Manufacturing	25.7%	5.4%
				325188: All Other Basic Inorganic Chemical Manufacturing	58.3%	2.9%
				325192: Cyclic Crude and Intermediate Manufacturing	102.2%	1.2%
				325199: All Other Basic Organic Chemical Manufacturing	49.4%	1.3%
				325211: Plastics Material and Resin Manufacturing	37.8%	1.0%
				325212: Synthetic Rubber Manufacturing	56.8%	1.0%
				325221: Cellulosic Organic Fiber Manufacturing	90.5%	1.9%
				325222: Noncellulosic Organic Fiber Manufacturing	40.3%	2.2%
				325311: Nitrogenous Fertilizer Manufacturing	83.0%	18.5%
Pulp, Paper, and Print	11.7%	0.7%	24.3%	322110: Pulp Mills	90.3%	1.8%
				322121: Paper (except Newsprint) Mills	17.1%	1.9%
				322122: Newsprint Mills	67.7%	5.1%
				322130: Paperboard Mills	25.1%	2.9%
Nonmetallic Minerals	19.7%	2.7%	30.8%	327111: Vitreous China Plumbing Fixture and China and Earthenware Bathroom Accessories Mfg.	59.9%	1.8%
				327112: Vitreous China, Fine Earthenware, and Other Pottery Product Manufacturing	94.0%	1.4%

Table 2. (Continued)

GTAP Energy-Intensive Trade-Exposed Sector	GTAP Sector's Trade Intensity[1]	GTAP Sector's GHG Intensity[1]	Estimate of "Presumptively Eligible" Industries' Share of Corresponding GTAP Sector's Output	"Presumptively Eligible" Industries in Corresponding GTAP Sector[2]	Industry's Trade Intensity[1]	Industry's GHG Intensity[1]
				327113: Porcelain Electrical Supply Manufacturing	40.7%	1.3%
				327122: Ceramic Wall and Floor Tile Manufacturing	69.0%	2.4%
				327123: Other Structural Clay Product Manufacturing	26.9%	3.5%
				327125: Nonclay Refractory Manufacturing	45.5%	1.7%
				327211: Flat Glass Manufacturing	51.1%	2.5%
				327212: Other Pressed and Blown Glass and Glassware Manufacturing	58.3%	3.7%
				327213: Glass Container Manufacturing	20.9%	2.4%
				327310: Cement Manufacturing	18.6%	15.9%
				327410: Lime Manufacturing	3.7%	33.0%
				327992: Ground or Treated Mineral and Earth Manufacturing	16.8%	3.0%
				327993: Mineral Wool Manufacturing	18.1%	1.5%
Iron and Steel	21.9%	2.3%	63.6%	331111: Iron and Steel Mills	33.3%	2.6%
				331112: Electrometallurgical Ferroalloy Product Manufacturing	77.0%	6.1%
				331210: Iron and Steel Pipe and Tube Manufacturing from Purchased Steel	N/A	0.6%
				331511: Iron Foundries	15.4%	1.6%
Nonferrous Metals	34.3%	1.4%	22.9%	331311: Alumina Refining	69.7%	9.0%
				331312: Primary Aluminum Production (GHG intensity reflected in Alumina Refining)	64.1%	
				331411: Primary Smelting and Refining of Copper	54.6%	0.3%
				331419: Primary Smelting and Refining of Nonferrous Metal (except Copper and Aluminum)	135.4%	1.9%

Notes:

1. Trade and GHG intensities are calculated in the same manner as in Table 1.
2. Nine of the "presumptively eligible" industries fall within other GTAP sectors that are not considered energy-intensive trade-exposed sectors in this modeling based on the energy and trade intensity of the vast majority of economic activity in those sectors. These nine industries account for less than eight percent of the total emissions and shipments of all "presumptively eligible" industries.

1. Effectiveness of LDC Allocations and Output-Based Allocations in Mitigating a Cap-and-Trade Program's Impacts on Industries' Domestic Production Costs

Figure 14 depicts the projected change in marginal production costs in each energy-intensive trade-exposed sector resulting from a cap-and-trade program with a $20 allowance price under three different policy scenarios. The dark gray bar on the left of each cluster shows the change in production costs without LDC and output-based allowance allocations. Differences between the changes in production costs depicted in Figure 14 and the GHG intensity measures presented in Table 2 reflect the indirect effects on production costs arising from the cap-and-trade program's broader impacts on the economy. In some cases, these indirect effects lead to greater increases in production costs than suggested by the sector's GHG intensity alone. For example, in addition to experiencing increased energy expenditures, industries that rely on other energy-intensive industries for intermediate inputs, such as cement or steel, may also see their production costs rise from increased costs for these intermediate inputs. In other cases, indirect effects — such as changes in underlying fuel prices or reduced economy-wide demand for key raw materials — can lead to lesser increases in production costs than suggested by a sector's GHG intensity alone. Note that, in this scenario, there is considerable variation in the cap-and-trade program's impacts on the five sectors' production costs, indicative of — though less substantial than — the variation that can also be expected across the more narrowly defined "presumptively eligible" industries, as shown in Figure 5.

The gray bar in the middle of each cluster shows how the introduction of LDC allocation mitigates the increase in production costs experienced by each sector.[50] Note that, while the LDC allocation mitigates impacts on the production costs of all energy-intensive trade-exposed industries, the degree of mitigation varies dramatically, dampening the increase in costs by more than half for the nonferrous metals sector and by just one-tenth for the nonmetallic minerals sector. This variation reflects the different contributions of electricity consumption to each sector's total direct and indirect emissions.

The light gray bar on the right of each cluster shows the change in production costs with both LDC allocations and output-based allocations. As this bar indicates, together, the LDC allocations and output-based allocations almost fully offset the cap-and-trade program's impacts on marginal production costs for three sectors, and more than fully offset those impacts for the two remaining sectors.

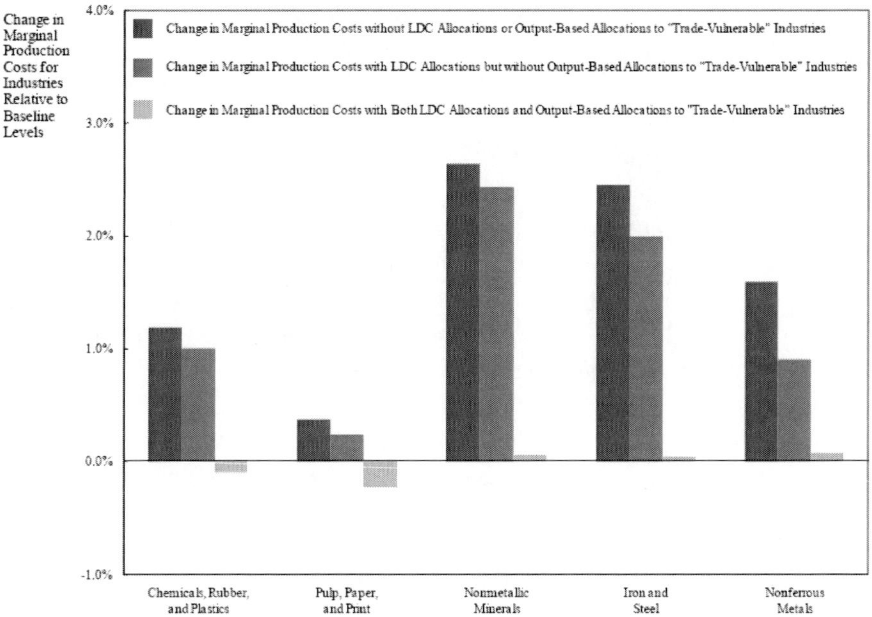

Figure 14. Effect of Domestic Cap-and-Trade Program on Marginal Production Costs of Energy-Intensive Trade-Exposed Industries without and with Allocations to Local Distribution Companies and Output-Based Allocations to "Trade-Vulnerable" Industries.

Some have suggested that the LDC allocations and output-based rebates will be insufficient to fully mitigate the increase in production cost borne by energy-intensive trade exposed industries even if the output-based rebates are provided at a rate equal to each sector's average emission-intensity. The rationale given for this argument is that these allocations will not directly address the cap-and-trade program's *indirect* impacts on industries' production costs, such as through effects on underlying fuel prices (*e.g.*, driven by a change in overall demand for natural gas and other fuels) or on the prices of key inputs other than energy.

Yet, as Figure 14 indicates, together, the LDC allocations and output-based rebates can, in fact, fully — and potentially more than fully — mitigate the increase in production costs borne by energy-intensive trade-exposed industries, and the associated competitiveness impacts, even after accounting for the program's indirect effects. At least two factors explain this conclusion. First, while some indirect effects of a cap-and-trade program can increase its effect on the production costs of affected industry, others, such as reductions in the nationwide demand for certain fossil fuels, can dampen a program's impacts on these industries' production costs. Second, in addition to the direct benefits that these allocations provide to recipients, they also provide indirect benefits, such as through their effects on the prices that industries face for various emission-intensive raw materials (*e.g.*, steel). Of course, under alternative assumptions regarding the factors that drive a cap-and-trade program's broader economic effects, such as its impacts on underlying fuel prices, modeling may find that the output-based allocations do not fully offset the cap-and-trade program's impacts on affected industries' production costs. But, regardless, output-based allocations at a rate equal to the average emission intensity of an industry's production will clearly mitigate the vast majority of impacts on manufacturers.

It should also be noted that, while the rightmost gray bars depict the combined effect of both the LDC allocations and the output-based rebates, it would be possible to achieve the same effect on energy-intensive industries' production costs through output-based rebates alone if those rebates were provided at a rate that addressed all indirect emissions, and not just the residual emissions left unaddressed by the LDC allocations. This would circumvent the potentially challenging task of evaluating the effective benefit that firms receive from LDC allocations and deducting this benefit from the output-based rebates. Indeed, H.R. 2454 is written so that the output-based rebates will increase to the extent the LDC allocation is reduced, though greater reliance on these rebates to cover firms' indirect emissions may require adjusting the total number of allowances set aside for output-based rebates, as can be seen in Figure 13.

2. Effectiveness of LDC Allocations and Output-Based Allocations in Mitigating a Cap-and-Trade Program's Impacts on Changes in Trade Flows that Can Give Rise to Leakage

While the previous discussion showed that H.R. 2454's allocations to LDCs and "trade-vulnerable" industries can fully mitigate a cap-and-trade program's impact on the domestic production costs of energy-intensive trade-exposed industries, we now examine how well these allocations mitigate impacts on trade flows that give rise to leakage. Figure 15 summarizes the effect of different combinations of U.S. and developed country cap-and-trade and allocation policies on U.S. net imports of the goods produced by energy-intensive trade-exposed industries. In this figure, changes in net imports are expressed as a share of domestic U.S. production. Each of the five industries is shown as a separate cluster of bars, with bars in each cluster representing different assumptions about U.S. and developed country policies. As a starting point for understanding impacts, the leftmost (black) bar in each cluster shows the impact of the United States acting alone without any mitigating measures. While this scenario is not realistic — given that, among other considerations, the European Union has already adopted a cap-and-trade program — this is the scenario that is examined in many prior analyses of competitiveness impacts of a domestic cap-and-trade program.

Consistent with other studies of these industries discussed at the beginning of this section, Figure 15 shows an increase in net imports on the order of 1 to 2 percent of total domestic production if other developed countries do not take action to reduce their emissions and if no domestic measures are adopted to mitigate the cap-and-trade program's competitiveness impacts. In reality, most developed countries are taking or plan to take action to reduce greenhouse gas emissions through cap-and-trade or other market-based mechanisms, and this significantly reduces the international competitiveness impacts of a domestic cap-and-trade program. The second bar in each cluster shows the impact on net imports with the more realistic assumption of action by all developed countries. For industries with significant baseline levels of net imports — non-metallic minerals, iron and steel, and nonferrous metals — action by other developed countries reduces the impacts of a U.S. cap-and-trade program on net imports by one-quarter to one-half.

We now turn to the effect of the allocations to LDCs and "trade-vulnerable" industries. In Figure 15 the right two bars in each cluster show what happens when we apply these mitigation measures in the case of action by all developed countries. We present two scenarios: one with LDC and output-based allocations in the United States alone (third bar), and another (the last bar) where other developed countries also employ output-based allocations for the "trade-vulnerable" industries at a rate equal to the average emissions-intensity of recipient industries' production. The two different scenarios reflect uncertainty regarding how the effects of existing and proposed allocation approaches in other developed countries, particularly in the European Union and Australia, will compare with those expected from the proposed LDC and "trade-vulnerable" industry allocations in H.R. 2454. For example, while the EU ETS is not tying its allocations to energy-intensive trade-exposed industries directly to their on-going production levels, plants will lose their otherwise fixed allocation if they shut down and allocations are also provided to new entrants. By attaching these conditions to allocations, the EU ETS allocations will have some — though not all — of the same effects as the output-based allocations proposed in H.R. 2454.

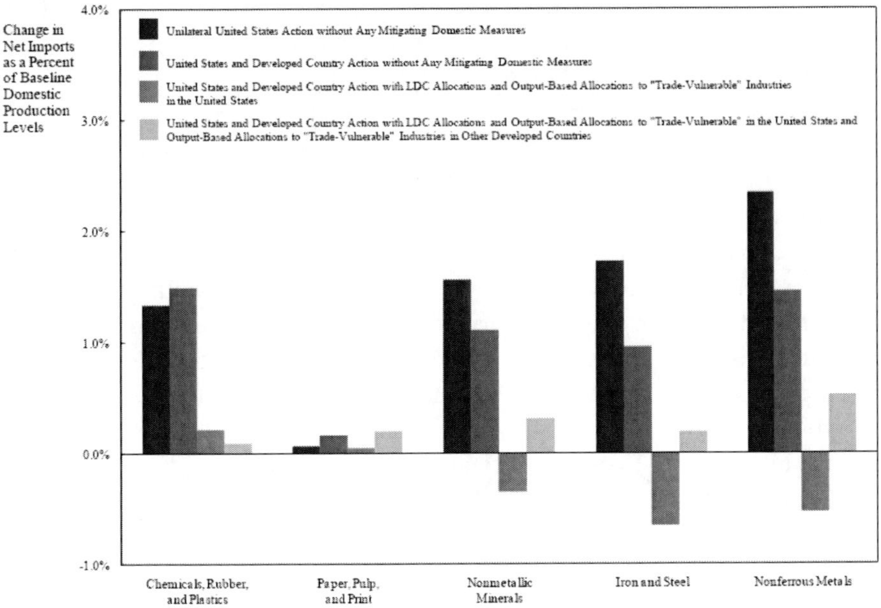

Figure 15. Changes in Total Net Imports of Energy-Intensive Goods under Alternative Policy Scenarios.

Overall, regardless of what steps other developed countries take to mitigate competitiveness impacts, the combination of allocations to LDCs and "trade-vulnerable" industries will dramatically reduce a domestic cap-and-trade program's effect on net imports within the energy-intensive trade-exposed industries.[51] As is depicted in Figure 15, if other developed countries do not adopt similar efforts to reduce the impact of climate policies on these industries, then U.S. net imports of these industries' goods could *decline* relative to a baseline in which no countries adopt climate policies. That is, when a U.S. cap-and-trade program includes these mitigation measures, net imports could be *lower* under the cap-and-trade program than they would be if neither the United States nor other developed countries adopted a cap-and-trade program. This result arises primarily from changes in trade flows between the United States and other developed countries that do not have similar policies to mitigate effects on energy-intensive trade-exposed industries.

When all developed countries adopt similar measures to reduce impacts on energy-intensive trade-exposed industries, net imports increase relative to a baseline in which no countries adopt climate policies. However, this increase is between just 0.1 and 0.3 percent of domestic production levels for four of the sectors, and reaches only 0.5 percent of domestic production levels in the iron and steel sector. These residual changes in net imports are a fraction of what would occur in the absence of the mitigation measures (*e.g.*, the left bar in each cluster in Figure 15) and are much smaller than the changes in imports and exports that these industries have experienced over the past decade in response to other economic factors (see Figure 8 and Figure 9).

As shown in Figure 14, the allocations to LDCs and "trade-vulnerable" industries essentially fully offset the cap-and-trade program's impact on the domestic production costs of energy-intensive trade-exposed industries. Consequently, the residual increases in U.S. net imports that occur when all developed countries adopt measures to mitigate competitiveness impacts result from the broader effects of climate policies on energy markets and the global economy, not from the direct impact of a cap-and-trade program's emission allowance requirement on the production costs of energy-intensive trade-exposed industries. One of these broader effects that likely drive the residual changes in U.S. net imports is the effect of climate policies on developed countries' demand for fossil fuels. By reducing this demand, the adoption of climate policies in developed countries can slightly lower global energy prices (*e.g.*, world oil prices). In turn, these changes in world energy prices can affect the production costs of energy-intensive trade-exposed industries in developed and developing countries alike, leading to modest shifts in trade flows between the United States and its developed and developing country trading partners.

3. Effectiveness of LDC Allocations and Output-Based Allocations in Promoting Real Reductions in Global Emissions from Energy-Intensive Trade-Exposed Industries

Finally, we turn to the effectiveness of allocations to LDCs and "trade-vulnerable" industries in reducing emission leakage from U.S. industry, and thereby enhancing a U.S. cap-and-trade program's ability to achieve real reductions in global emissions. The predominant concern regarding emission leakage associated with a cap-and-trade program's impact on the international competitiveness of U.S. industry is that such a program will lead to a shift in manufacturing activity to developing countries that do not yet have comparable greenhouse gas regulations. In light of this, whereas Figure 15 depicted changes in *total* U.S. net imports, Figure 16 depicts changes in U.S. net imports from developing countries under three alternative policy scenarios.

As was previously noted, other developed countries already have adopted — or likely will soon adopt — mandatory climate policies. Through the broader economic effects that it can have on the global economy and world energy markets, the adoption of climate policies in other countries can have implications for U.S. trade flows even if the United States does not adopt a cap-and-trade program. In light of this, to realistically assess the competitiveness effects of U.S. adoption of a cap-and-trade program, it is important to identify and differentiate those changes in trade flows that are attributable to U.S. policy choices, and those that are attributable to actions that other developed countries have already taken, or may soon take, to reduce their emissions.

To realistically assess the incremental effect of the United States' adoption of a cap-and trade program on its net imports from developing countries, the first bar in Figure 16 depicts how U.S. net imports from developing countries would change in response to the adoption of cap-and-trade programs *in other developed countries even if the United States itself does not adopt such a program.* Given the current state of, and outlook for, global action to address climate change, this scenario offers a good point of comparison in assessing leakage resulting from the U.S. adoption of a cap-and-trade program in this global context.

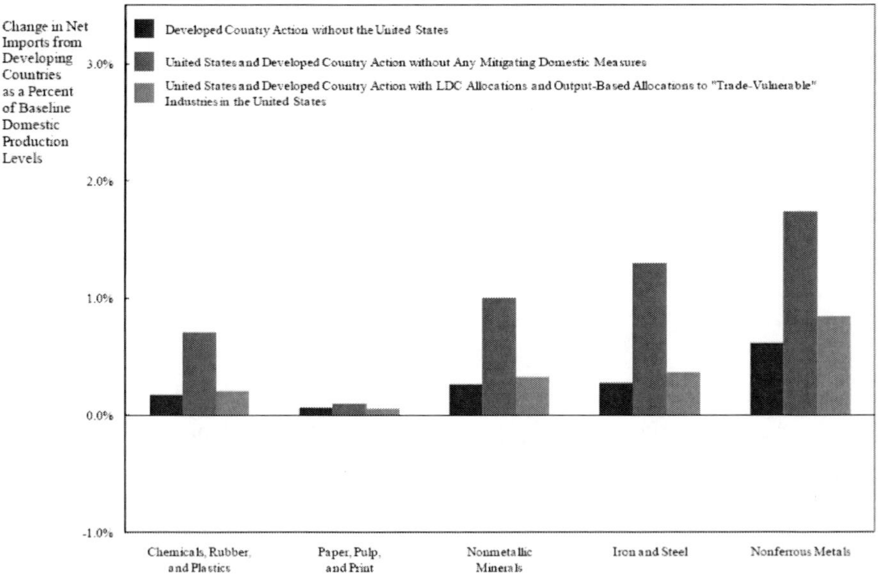

Figure 16. Effect of U.S. Adoption of a Cap-and-Trade Program on Net Imports of Energy-Intensive Goods from Developing Countries under Alternative Policy Scenarios.

The second bar then shows how U.S. net imports from developing countries would change if the United States joins other developed countries in adopting cap-and-trade programs — still without any mitigating policy measures to address competitiveness impacts.[52] As is made evident by a comparison of the first and second bar in Figure 16, when both the United States and other developed countries adopt cap-and-trade programs, only a portion of the resulting change in U.S. net imports from developing countries depicted in the second bar is attributable to U.S. adoption of a cap-and-trade program. Some portion of that change (the amount depicted in the first bar) would occur even if the United States did not adopt a cap-and trade program.

Finally, the third bar depicts how U.S. net imports from developing countries change in response to the adoption of cap-and-trade programs in the United States and other developed countries if the United States employs the allocations to LDCs and "trade-vulnerable" industries set out in H.R. 2454. While U.S. net imports from developing countries still increase in the last policy scenario, the allocations to LDCs and "trade-vulnerable" industries offset nearly all of the increase in net imports that can actually be attributed to the U.S. adoption of a cap-and-trade program (the third bar is essentially the same height as the first bar in each cluster). That is, if the United States adopts a cap-and-trade program that includes the allocations to LDCs and "trade-vulnerable" industries, net imports from developing countries in the energy-intensive trade-exposed industries (third bar) will be essentially the same as they would be if developed countries adopt cap-and-trade programs without United States action (first bar).

Figure 17 translates the changes in net imports from developing countries depicted in Figure 16 into associated emission leakage, comparing the level of this leakage with total emission reductions in each sector. For the reasons described above, Figure 17 evaluates the effects of a U.S. cap-and-trade program and the mitigating allowance allocations against a baseline in which the other developed countries act with or without the United States.[53] It then shows the emission reductions and emission leakage in each energy-intensive trade-exposed industry that arise from the U.S. adoption of a cap-and-trade program both without (the left-hand side) and with (the right-hand side) allocations to LDCs and "trade-vulnerable" industries.

We start with the estimated emission reductions against which leakage occurs. The emission reductions depicted by the columns in Figure 17 (which reflect both direct emissions and indirect emissions from electricity consumption) are decomposed into those reductions that arise from reductions in the emission intensity of domestic production and those that result from reductions in domestic production levels. In turn, the emission reductions arising from reductions in domestic production are further decomposed into those reductions in domestic production arising from reduced domestic consumption, and those arising from changes in net imports. Finally, the black line overlain on each column depicts emission leakage: the increase in emissions in developing countries arising from the cap-and-trade program's effects on U.S. net imports from developing countries.[54]

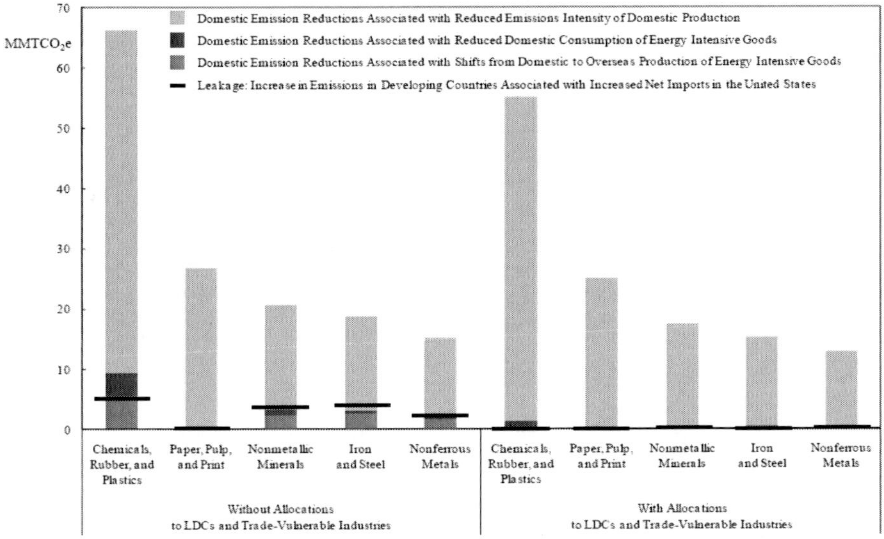

Note: To isolate the effects of U.S. adoption of a cap-and-trade program on domestic emission reductions and leakage, this chart depicts the *incremental* emission reductions and leakage that result from the U.S. adoption of a cap-and-trade program in the context of a policy scenario in which the United States and other developed countries adopt comparable climate policies. Specifically, the emission reductions and leakage depicted here reflect the change in emissions and leakage in a scenario in which the United States and other developed countries adopt comparable cap-and-trade programs, relative to the level of emissions and leakage that would prevail in a scenario in which other developed countries adopt such programs but the United States does not.

Figure 17. Estimated Emissions Reductions and Leakage from U.S. Energy-Intensive Trade-Exposed Industries without and with Allocations to LDCs and "Trade-Vulnerable" Industries.

Figure 17 shows that the overwhelming majority of emission reductions in the energy-intensive trade-exposed sectors come from reductions in the emission intensity of domestic production (e.g., from more efficient energy use and from fuel switching). This is noteworthy because, as Figure 17 depicts, reductions in the emission intensity of domestic production will occur both with and without the output-based allocations to "trade-vulnerable" industries.[55] By contrast, emission reductions owing to changes in the level of domestic production are relatively small regardless of the allocation approach that is adopted. Absent any mitigating allocations, the left panel shows that leakage (the black line) is a varying share of these reductions, owing to the degree to which lower domestic production reflects increased net imports, the degree to which that change in net imports reflects trade with unregulated developing countries, and the emission intensity of developing country production vis-à-vis U.S. production. Even without the mitigating allocation measures, at just 15 MMTCO2e across all five industries, the overall level of emission leakage associated with shifts in manufacturing activity from the United States to developing countries is quite small relative to the overall level of emission reductions that will be achieved under a domestic cap-and-trade program.

As the right-hand panel in Figure 17 depicts, however, the combination of allocations to LDCs and "trade-vulnerable" industries eliminate effectively all of the emission leakage that would otherwise arise from shifts in manufacturing activity to developing countries following the adoption of a U.S. cap-and-trade program. Of course, while these allocations preserve reductions in the emission-intensity of production that a cap-and-trade program can achieve, they forego opportunities to reduce emissions through reduced domestic consumption of emission-intensive goods, as is evident from the reduction in the size of the dark gray middle component of each column in moving from the left to the right panel. Nonetheless, this modeling suggests that, on average, for every ton of emission reductions from the reduced domestic consumption of emissions-intensive goods that these allocations forego, they avoid three tons of emission leakage.

VIII. CAVEATS AND NEXT STEPS

While this analysis has found that the allocation provisions in H.R. 2454 can effectively mitigate the adverse impacts that a cap-and-trade program might otherwise have on the international competitiveness of — and associated leakage from — industry, a key concern of many stakeholders is that impacts may be more pronounced for more narrowly defined industries that have even higher emission intensities and/or trade sensitivities. However, the ability of output-based allocations to mitigate these impacts is relatively robust, hinging primarily on the ability of the allocation level to match an industry's emission level.

Indeed, studies examining output-based allocations in the context of more narrowly defined industries in the European Union have come to similar conclusions (see Section V).

As more detailed economic models are adjusted to model climate policies, it may be possible to examine effects on more narrowly defined industries. Effects on individual firms will also vary, depending on their emission intensity relative to the industry average. Survey data could be used to look at intra-industry variation.

While this paper demonstrates that, on a theoretical basis, output-based allocations at the levels proposed in H.R. 2454 can effectively eliminate the competitiveness impacts and associated leakage resulting from the uneven imposition of climate policies throughout the world, the challenges of implementation have not been fully considered here. The conclusions of this paper rest on at least three key assumptions:

- That, in a timely manner, EPA will be able to obtain data from facilities on output levels, electricity use, and emissions associated with electricity use (in addition to data already planned via the Mandatory Reporting Rule);
- That such data can be generated at a sufficiently disaggregated level for EPA to develop meaningful benchmarks for output-based allocations; and
- That LDC allocations will indeed lead to the intended distribution of benefits to industrial firms across the country.

Though we believe these implementation challenges are surmountable, each requires careful consideration. Further discussion is beyond the scope of this paper.

The analysis prepared in this report relies heavily on the Manufacturing Energy Consumption Survey (MECS) that is fielded every four years by the Energy Information Administration.

Increasing the frequency and expanding the scope of the MECS could greatly improve future analysis of this sort. Currently, some of the "presumptively eligible" industries are not covered in the MECS, so details about their energy use are not known.

Furthermore, increasing the frequency of the survey to every two or three years together with enlarging the sample size would help to better understand how industry energy usage patterns react to changes that might be stimulated by regulations associated with climate change legislation. It would also greatly improve the accuracy of the estimated data.

This report represents a first step in the Administration's engagement on these issues and more work will need to be done both to continue to improve assessments of competitiveness impacts at more disaggregated levels, and to address the various implementation challenges presented by output-based allocations.

APPENDIX A: ACRONYMS

ADAGE	Applied Dynamic Analysis of the Global Economy: a dynamic computable general equilibrium model capable of investigating economic policies at the international and U.S. national, regional, and state levels
BAU	Business as usual
CO2	Carbon dioxide
CO2e	Carbon dioxide equivalent
EIA	Energy Information Administration
EPA	Environmental Protection Agency
EU ETS	European Union Emission Trading Scheme
GHG	Greenhouse gas
GTAP	Global Trade Analysis Project: a global trade model with a database that contains bilateral trade information for over 40 countries for 50 sectors.
LDC	Local distribution company
MECS	Manufacturing Energy Consumption Survey
MMTCO2e	Million metric tons of carbon dioxide equivalent
MWh	Megawatt-hour
NAICS	North American Industry Classification System
U.S. ITC	United States International Trade Commission

APPENDIX B: STAKEHOLDERS AND OTHER GROUPS CONSULTED IN PREPARING THIS REPORT

During the course of preparing this report, the Administration consulted with various stakeholders[†] and other groups, including:

Aluminum Association
American Chemistry Council
American Forest and Paper Association
American Iron and Steel Institute
American Materials Manufacturing Alliance
Blue Green Alliance
Center for Clean Air Policy
Compressed Gas Association
Energy-Intensive Manufacturers' Working Group on Greenhouse Gas Regulation
United Steelworkers
World Resources Institute

APPENDIX C: LETTER FROM SENATORS BAYH, SPECTER, STABENOW, MCCASKILL, AND BROWN

United States Senate
WASHINGTON, DC 20510
September 11, 2009

The Honorable Carol Browner
Assistant to the President for Energy and Climate
Change The White House

Washington, D.C. 20500

[†] Representatives of individual companies often accompanied representatives of trade associations in meetings with the Administration. Those individual companies are not listed here.

Dear Ms. Browner:

Climate change is one of the most far-reaching economic challenges of our times. With sensible, comprehensive policies, mitigating the worst of its impacts will help create vibrant new industries and jobs, and leave us with a cleaner and more secure world.

Efforts to contain costs for energy-intensive and trade-exposed (EITE) manufacturers, if not properly considered, can harm other industries and raise the costs of reducing emissions for the whole economy. We are writing therefore to request assistance in assessing the EITE provisions in Section 782(e) of the America Clean Energy and Security Act of 2009, H.R. 2454.

First, we are seeking an analysis with the following elements:

1) Determination of which industrial sectors of the economy are likely to be eligible for allowance rebates under the EITE provisions of the bill and how many direct and indirect emissions can be attributed to manufacturing by facilities within these industrial sectors.
2) Assessment of potential competiveness impacts to energy intensive and trade exposed manufacturers if the United States adopts a climate change policy to address greenhouse gas emissions. We recognize that many factors affect competiveness, thus we seek to focus the assessment specifically and exclusively on the potential foreign competition losses from cap and-trade adoption in the U.S. Such an assessment should make reasonable assumptions regarding the extent to which our trading partners have adopted comparable policies.
3) Identify additional data that, if provided, could improve the assessment of competitiveness impacts and allocation requirements, and greenhouse gas emissions from other countries. We ask that you proactively reach out to industrial stakeholders to determine if they would be willing to provide internal data that could improve the assessment of the impacts on their respective sectors.
4) Incorporate into the analysis other measures, including but not limited to those provisions in H.R. 2454 that may be beneficial to EITE industries, such as industrial energy efficiency incentives, allocations to Local Distribution Companies, and carbon capture and sequestration incentives.

The Honorable Carol Browner

We hope the analysis would also take into consideration results from economic models that capture international trade flows, such as those used by the U.S. Environmental Protection Agency, the Department of Energy, the Energy Information Administration, and the U.S. International Trade Commission. The Departments of Commerce and Treasury and the Office of the U.S. Trade Representative may also have analytical expertise that should be brought to bear on the overall analysis.

Second, industrial constituents have approached us with concerns that the EITE-related datasets referenced in H.R. 2454 contain errors that would result in them being treated contrary to the legislation's intent with regard to eligibility or future allocations. Some industries also claim that despite the output-based rebate, their sectors will still be substantially impacted by the cap-and trade proposal as a whole. We, however, do not possess the technical capacity to evaluate these claims. Thus, we request that you create a process by which we can direct industrial firms to technical experts within the Administration to evaluate their concerns.

Third, we would like to request a proactive strategy to brief all industry stakeholders on the study, so they can gain a fuller understanding of all the likely provisions of a cap-and-trade bill that may help their industries.

We do not want this request to delay our leadership's desired schedule on climate change legislation. Given the scarcity of time, we therefore see an immediate need to begin work on this request. Schedules permitting, we respectfully request a meeting next week with the appropriate Administration officials and staff to develop a work plan for the points listed above.

Sincerely,

Sen. Evan Bayh

Sen. Arlen Specter

Sen, Debbie Stabenow

Sen. Claire McCaskill

Sen. Sherrod Brown

The Honorable Carol Browner

Cc: The Honorable Shara Aranoff, Chairman, U.S. International Trade Commission

The Honorable Steven Chu, Secretary of Energy

The Honorable Timothy Geithner, Secretary of the Treasury

The Honorable Lisa Jackson, Administrator, U.S. Environmental Protection Agency

The Honorable Ron Kirk, Ambassador, U.S. Trade Representative

The Honorable Gary Locke, Secretary of Commerce

The Honorable Richard Newell, Administrator, Energy Information Administration

End Notes

[1] This report focuses on one significant source of emission leakage whereby the adoption of domestic greenhouse gas regulations leads to shifts in emission-intensive manufacturing activity and associated emissions to countries without comparable regulations. The report focuses on this type of leakage because of the nature of the Senators' request and because it can be prevented through careful domestic policy design. There are, however, other ways in which the adoption of domestic regulations can lead to increased emissions abroad that cannot be controlled through domestic policy design. For example, by reducing U.S. demand for oil, and thereby world oil prices, U.S. greenhouse gas regulations can lead to increased consumption of oil (and increased emissions) abroad, which would partly offset the reductions achieved at home.

[2] While this report focuses on H.R. 2454, reflecting the request from the Senators, the provisions in H.R. 2454 are very similar to those in the Clean Energy Job and American Power Act of 2009 (S. 1733).

[3] In addition to drawing on publicly available data, the Administration has met with various stakeholder groups, including representatives of energy-intensive industries, and has requested their assistance going forward in obtaining additional data that is not yet available in the public domain, but that would improve assessments of affected industries and competitiveness impacts. While some stakeholders have provided the Administration with data and analyses that can help better inform various aspects of the discussion about competitiveness impacts, what has been provided to date does not affect the specific analyses and estimates presented in this report.

[4] The criteria established in S. 1733 are generally consistent with those in H.R. 2454.

[5] An industry's energy intensity is defined as its energy expenditures as a share of the value of its domestic production. An industry's greenhouse gas intensity is defined as its total greenhouse gas emissions (including indirect emissions from electricity consumption) times $20 per ton of emissions, divided by the value of the industry's domestic production. An industry's trade intensity is defined as the combined value of its exports and imports as a share of the value of its domestic production and imports. This paragraph describes the

general criteria established for identifying "presumptively eligible" industries and does not address additional, more detailed eligibility considerations set forth in H.R. 2454. However, the preliminary eligibility assessment presented in Table 1 reflects the application of all the criteria and considerations set forth in H.R. 2454.

[6] NAICS is the standard classification system used by federal statistical agencies in classifying business establishments for the purpose of collecting, analyzing, and publishing statistical data related to the U.S. business economy. The most detailed public data on the production and operation of manufacturing facilities is provided at a six-digit industry classification level. Within manufacturing, there are nearly 500 six-digit sectors. Classifications with fewer digits (*e.g.*, at the four-digit level) aggregate these six-digit sectors into more broadly defined sectors.

[7] For example, relevant data from the 2007 Economic Census and the Energy Information Administration's 2006 Manufacturing Energy Consumption Survey were released over the summer.

[8] Carbon dioxide equivalent is a measure of emissions that expresses non-CO2 greenhouse gas emissions in terms of the number of tons of CO2 that would have the same global warming potential over a given timeframe as those nonCO2 emissions. This report assumes a 100-year timeframe, consistent with H.R. 2454. While the vast majority of the emission estimates are based on 2006 data, estimates for ten of the "presumptively eligible" industries, representing 6% of total emissions from "presumptively eligible" industries, are based on 2007 data.

[9] The MECS offers some of the best available data on individual industries' energy-related emissions, but it does not provide data for all "presumptively eligible" industries. It was possible to use data from the 2006 MECS to estimate direct combustion emissions and indirect electricity emissions for 25 of the "presumptively eligible" industries, which account for 90% of the total estimated direct combustion and indirect electricity emissions of all "presumptively eligible" industries.

[10] S. 1733 also provides these options.

[11] While natural gas for "non-emissive" uses (*e.g.*, as a feedstock) would not be subject to the requirement to submit allowances for emissions under H.R. 2454, there could still be competitiveness impacts on industries that use natural gas as a feedstock if the demand for and price of natural gas is affected by the cap-and-trade program.

[12] Environmental Protection Agency (2009), *EPA Analysis of the American Clean Energy and Security Act of 2009.*

[13] Energy Information Administration (2009), *Energy Market and Economic Impacts of H.R. 2454, the American Clean Energy and Security Act of 2009*. Because wellhead prices are lower than delivered prices, these percentage changes are not directly comparable with the above estimates from the EPA analysis.

[14] The one scenario in which prices were projected to rise beyond these ranges was an extreme, worst case scenario in which, through 2030, no international emission reduction offsets would be available and several key clean energy technologies (e.g., nuclear, and carbon capture and storage) could not be expanded beyond business-as-usual levels.

[15] S. Paltsev, J. Reilly, H. Jacoby, and J. Morris (2009), *The Cost of Climate Policy in the United States*, MIT Joint Program on the Science and Policy of Global Change, Report No. 173, updated in September 2009.

[16] Resources and reports on industrial energy efficiency can be found at http://www.energystar.gov/index.cfm?c= industry.bus_industry_info_center, and http:// www.eere.energy.gov/.

[17] U.S. Department of Energy, Office of Energy Efficiency and Renewable Energy (2005), *Steel Industry Marginal Opportunity Study*, prepared by Energetics, Inc., pp. 3, 7.

[18] Note that while both integrated and electric arc steelmaking are combined into a single six-digit NAICS code for the purposes of industrial classification and for the determination of eligibility for allowance rebates under H.R. 2454, they are considered separate sectors within H.R. 2454 when determining the rate of allocation provided per unit of output. See Section VI.A.1 for further discussion of how these per-unit allocation rates are determined.

[19] J. Ederington, A. Levinson, and J. Minier (2005), "Footloose and Pollution-Free," *Review of Economics and Statistics* 87: 92-99; and J. Aldy and W. Pizer (2009), *The Competitiveness Impacts of Climate Change Mitigation Policies,* Pew Center on Global Climate Change. http://www.pewclimate.org/international/CompetitivenessImpacts.

[20] JC. Hourcade, D. Demailly, K. Neuhoff, and M. Sato (2007), *Climate Strategies Report: Differentiation and Dynamics of EU ETS Industrial Competitiveness Impacts.*

[21] Ederington et al. (2005).

[22] Hourcade et al. (2007).

[23] *Ibid.*

[24] *Ibid.*

[25] W. Antweiler, B.R. Copeland, and M.S. Taylor (2001), "Is Free Trade Good for the Environment?" *American Economic Review* 91(4): 877-908.

[26] Hourcade et al. (2007).

[27] Ederington et al. (2005).

[28] Note that these statistics represent aggregated values for each broad sector, and some "presumptively eligible" industries' trade patterns could differ significantly from those of the aggregated sector within which they fall.

[29] M. Grubb, T. Brewer, M. Sato, R. Heilmayr, and D. Fazekas (2009), *Climate Policy and Industrial Competitiveness: Ten Insights from Europe on the EU Emissions Trading System,* German Marshall Fund. Available at: http://www.gmfus.org/publications/article.cfm?id=640&parent_type=P.

[30] Carbon Trust (2008), *EU ETS impacts on profitability and trade, a sector by sector analysis.* Available at: http://www.carbontrust.co.uk/Publications/publicationdetail.htm?productid=CTC728&metaNoCache=1.

[31] D. Demailly and P. Quirion (2006), "CO2 abatement, competitiveness and leakage in the European cement industry under the EU ETS: Grandfathering vs. output-based allocation," *Climate Policy* 6: 93-113.

[32] Carbon Trust, 2008.

[33] See R. Baron, R. Lacombe, P. Quirion, J. Reinaud, N. Walker, and R. Trotignon (forthcoming), "Competitiveness under the EU Emissions Trading Scheme," in A. Denny Ellerman, Frank J. Convery, and Christian De Perthuis, *Pricing Carbon: The European Union Emissions Trading Scheme*, Cambridge: Cambridge University Press. Also available at: http://www.aprec.fr/documents/wpcompetitiveness.pdf.

[34] See, for example, G. Cook (2009), "Climate Change and the Cement Industry," Working Paper, Climate Strategies, September 18. Available at http://www.climatestrategies.org/our-reports/category/32/222.html. See also J. Reinaud (2008), *Issues behind Competitiveness and Carbon Leakage, Focus on Heavy Industry*, International Energy Agency (IEA) Information Paper, OECD/IEA, October.

[35] H.R. 2454 establishes an emissions eligibility threshold of 25,000 tons CO2e annually for facilities within certain industrial sectors; entities within these sectors that fall below the threshold do not face a compliance obligation for their direct emissions.

[36] A numerical illustration may be useful. Suppose that Steely Steel Corp. produced 90,000 tons of steel two years ago and 110,000 tons of steel last year for an average output of 100,000 tons per year. If the steel sector's average direct GHG emissions per unit of steel over the last four years was 1 ton of CO2e emissions per ton of steel, Steely Steel Corp. would receive allowance rebates for 100,000 tons of direct emissions (100,000 tons of steel x 1 ton CO2e/ton steel). In addition, if the sector-average electricity use for the steel sector was 0.4 megawatt-hours (MWh) per ton of steel, and if the emission intensity of the electricity purchased by Steely Steel Corp. over the last year was 0.5 tons of CO2e emissions per MWh, Steely Steel Corp. could receive allowance rebates for up to 20,000 tons of indirect emissions (100,000 tons steel x 0.4 MWh/ton steel x 0.5 tons CO2e/MWh). As noted in the paragraph above, however, this allowance rebate for indirect emissions would be adjusted to account for any mitigation of electricity rate impacts through allocations to electricity LDCs.

[37] By adjusting the output-based allowance rebates for indirect emissions (associated with electricity purchases) to account for the value of allowances received through the LDC allocation, H.R. 2454 is designed to compensate sectors for the emissions associated with electricity purchases, but avoid doing so twice through output-based allocations and LDC allocations.

[38] Note that the non-emissive use of natural gas (*e.g.*, as a feedstock) will not create a compliance obligation for entities whose direct emissions are regulated. Further, if allowances have been previously retired for the emissions that would have resulted from the combustion of natural gas — as might happen if the natural gas was sold by an LDC to a non-covered entity — but this natural gas is used instead as a feedstock, a compensatory allowance would be issued to make up for the allowances that were retired.

[39] For example, see J. Sijm, K. Neuhoff, and Y. Chen (2006), "CO2 Cost Pass-Through and Windfall Profits in the Power Sector," *Climate Policy,* 6(1): 49-72.

[40] In particular, if allowances are allocated based on historical emissions or some other factor unrelated to on-going production levels, then an unregulated firm operating in a competitive market will make pricing and production decisions as if it had to purchase every allowance that it must use to cover its emissions. Even though the firm benefits from the allocation, because the level of the allocation is unaffected by future production decisions, it is equivalent to a one time gain that does not affect the firm's on-going production costs. If the firm generates emissions with each unit of output that it produces, to cover those emissions, it still has to surrender valuable allowances that it could have otherwise sold. As a result, its pricing decisions will reflect the value of those allowances despite the fact that it may have received the allowances for free.

[41] Estimates of the extent that the LDC allocations will raise the cap-and-trade program's cost vary significantly depending on several assumptions, including assumptions about how significant the foregone opportunities for emission reductions from reduced electricity use are, and assumptions about the cost of alternative means of reducing emissions, including the use of international emission reduction offsets. One analysis by Dallas Burtraw and others at Resources for the Future found that LDC allocations raise a cap-and-trade program's allowance prices — a proxy for the program's cost — by nearly 30 percent and that more than two-thirds of that increase is attributable to the effect of LDC allocations on incentives for reducing electricity use in the commercial and industrial sectors. By contrast, while EPA's analysis of H.R. 2454 did not directly address the impact of LDC allocations on allowance prices, the results from the different scenarios it examined imply that LDC allocations would raise allowance prices by only 2 percent. See R. Sweeney, J. Blonz, and

D. Burtraw (2009), "The Effects on Households of Allocation to Electricity Local Distribution Companies," June 5; and Environmental Protection Agency (2009), *EPA Analysis of the American Clean Energy and Security Act of 2009*, June 23.

[42] Aldy and Pizer (2009); and M. Ho, R. Morgenstern, and J.S. Shih (2008), "Impact of Carbon Price Policies on U.S. Industry," Discussion Paper DP-08-37, Resources for the Future.

[43] We use the term "short-run" to describe analyses that assume capital is fixed. "Long-run" studies are those that allow for capital turnover, accounting for how industries will adjust to a carbon price through eventual investments in new low-carbon equipment and processes.

[44] The authors note that they do not account for process emissions (*i.e.*, those not associated with fuel combustion), so their estimates likely understate impacts on cement manufacturers and others with significant process emissions.

[45] GTAP (Global Trade Analysis Project) is a global network of researchers who conduct quantitative analysis of international economic policy issues, especially trade policy. They cooperate to produce a consistent global economic database, covering many sectors and all parts of the world.

[46] See C. Fischer and A. Fox (2007), "Output-Based Allocation of Emissions Permits for Mitigating Tax and Trade Interactions," *Land Economics*, 83: 575-599.

[47] The international economic data used in ADAGE is drawn from the GTAP database.

[48] For example, the model does not currently have the capability to account for the availability of offsets, or to model how opportunities to bank and borrow emission allowances would affect allowance prices in any given year.

[49] As was noted in Section VI.A.3, considering both the output-based allocations to "trade-vulnerable" industries and the indirect benefit those industries would receive from the LDC allocations, the number of allowances set aside for these allocations exceed the 2006 emissions of the "presumptively eligible" industries. Whether there are, in fact, sufficient allowances to cover eligible industries emissions in the future will depend on whether additional industries are found to be eligible, and on future trends in the eligible industries' emissions. While future increases in output would tend to increase their emissions, continued reductions in the emission intensity of their production would have the opposite effect. Moreover, the emissions estimates depicted in Figure 13 reflect 2006 levels. Any consideration of the potential for increased emissions in the future resulting from increased production levels would therefore need to consider the effects of the recent economic recession on production levels.

[50] Consistent with the discussion in Section VI.A.2, the LDC allocation is modeled as covering 80 percent of the indirect emissions associated with electricity consumption.

[51] While, in the absence of any mitigating policy measures, a cap-and-trade program's impact on more narrowly-defined industries can be significantly greater than the aggregated impacts shown in Figure 15, allocations to LDCs and "trade-vulnerable" industries can be equally effective in mitigating the vast majority, if not all, of these greater impacts. This point has been demonstrated in prior analyses of competitiveness impacts in the EU ETS (see Figure 12).

[52] This bar reflects the same scenario as was presented in the second bar of each cluster in Figure 15. Thus, the second bar in each cluster in Figure 16 shows the portion of the change in total net imports in the second bar in Figure 15 that is associated with changes in net imports from developing countries.

[53] That is, the emission reductions and leakage depicted here reflect the change in emissions and leakage in a scenario in which the United States and other developed countries adopt comparable cap-and-trade programs, relative to the level of emissions and leakage that

would result from a scenario in which other developed countries adopt such programs but the United States does not.

[54] There are two reasons why emission leakage, the increase in developing country emissions resulting from increased U.S. net imports from developing countries (depicted by the black line), can differ from the reduction in U.S. emissions associated with changes in U.S. net imports (depicted by the lowest component of each column in Figure 17). First, the reduction in U.S. emissions associated with changes in U.S. net imports reflects changes in total net imports, not just net imports from developing countries. Second, even if one were to isolate the change in U.S. emissions associated with changes in U.S. net imports from developing countries, this will differ from the associated emission leakage if the increase in net imports is from a country where the emission intensity of production differs from that in the United States.

[55] However, as noted previously, the LDC allocations can reduce incentives that a cap-and-trade program would otherwise create for firms to reduce the electricity intensity of their production.

INDEX

A

abatement, 139
access, 92
accounting, 82, 87, 123, 141
adaptation, 34
adjustment, 21, 23, 24, 29, 30, 78, 102, 109
adverse effects, 33
agencies, 138
aggregation, 118
agricultural sector, 40, 47
agriculture, 10, 14, 16, 41, 45, 47, 50
airports, 62
American Recovery and Reinvestment Act, 68, 70, 109, 115
American Recovery and Reinvestment Act of 2009, 68, 70, 109, 115
antidumping, 21, 22, 36
assessment, ix, 66, 73, 135, 138
atmosphere, 29
authority, 22
automobiles, 4, 10, 43

B

banking, 107, 115
base, 23, 102
batteries, 110
Belarus, 62
benchmarks, 69, 132
benefits, 10, 33, 102, 104, 109, 124, 132, 140
biomass, 110
bonds, 108, 110
border adjustments, vii, 1, 4, 5, 15, 18, 19, 20, 21, 22, 23, 24, 25, 26, 27, 28, 29, 30, 31, 32, 35, 36, 37, 43, 55
bounds, 26
burn, 6
businesses, 59

C

cap-and-trade program, vii, viii, 1, 3, 4, 5, 6, 7, 10, 14, 16, 18, 19, 24, 25, 27, 28, 31, 32, 33, 34, 39, 40, 41, 43, 51, 55, 59, 61, 62, 65, 67, 69, 70, 71, 82, 84, 85, 86, 87, 89, 91, 93, 98, 100, 101, 102, 103, 104, 107, 111, 112, 113, 114, 115, 116, 118, 122, 123, 124, 125, 126, 127, 128, 129, 130, 131, 138, 140, 141, 142
capital intensive, 92
carbon, vii, 1, 2, 3, 4, 5, 6, 7, 9, 10, 13, 14, 17, 18, 19, 24, 25, 27, 28, 31, 32, 33, 34, 39, 40, 41, 42, 43, 44, 46, 47, 51, 52, 53, 54, 55, 56, 57, 59, 60, 61, 62, 66, 78, 82, 87, 89, 91, 107, 108, 109, 112, 133, 135, 138, 141
carbon dioxide, vii, 1, 6, 13, 17, 39, 40, 42, 44, 46, 53, 54, 66, 78, 82, 91, 109, 133

carbon emissions, 27, 108
carbon leakage, vii, 2, 9, 10, 18, 34, 39, 40, 41, 42, 44, 47, 52, 55, 56
case law, 30
categorization, 78
census, 48, 72, 73, 77, 78, 82, 84, 94, 95, 105, 138
challenges, 5, 25, 132, 133, 135
chemical industry, 16
chemicals, 3, 6, 14, 16, 39, 41, 45, 48, 50, 52, 56, 66, 67, 69, 73, 98, 108, 112, 117
China, 47, 58, 75, 81, 119
chlorine, 98
classes, 102
classification, 73, 138, 139
clean air, 29
clean energy, 68, 69, 70, 102, 108, 109, 115, 138
climate, viii, 1, 6, 29, 58, 59, 65, 66, 68, 69, 70, 82, 85, 87, 106, 107, 111, 112, 113, 114, 116, 126, 127, 128, 130, 132, 133, 135, 136
climate change, viii, 60, 66, 69, 70, 106, 113, 116, 128, 133, 135, 136
coal, viii, 6, 9, 10, 14, 16, 22, 39, 40, 42, 45, 50, 52, 56, 57, 59, 86, 87
coke, viii, 14, 39, 50
combined effect, 124
combustion, 6, 51, 71, 77, 100, 110, 138, 140, 141
commercial, 9, 41, 44, 103, 108, 140
commodity, 48
communities, 69
compensation, 14
competition, 4, 10, 13, 17, 18, 19, 34, 41, 42, 43, 44, 45, 47, 49, 50, 51, 52, 54, 61, 62, 91, 92, 93, 115, 135
competitive advantage, 8
competitive conditions, 91
competitiveness, viii, ix, 1, 3, 4, 5, 7, 9, 10, 15, 16, 18, 20, 21, 31, 34, 35, 43, 49, 53, 55, 56, 57, 61, 65, 66, 67, 68, 69, 70, 91, 93, 100, 103, 107, 111, 112, 113, 115, 116, 118, 123, 125, 126, 127, 128, 129, 132, 133, 135, 137, 138, 139, 141
competitors, 7, 14, 42, 49, 56, 70, 92, 102, 114, 118
compliance, 100, 139, 140
computer, 45, 47, 50
confidentiality, 72
congress, 1, 6, 14, 16, 35, 36, 47, 62
Congressional Budget Office, 1, 10, 13, 17, 34, 35, 39, 44, 46, 48, 53, 54, 61, 67, 115
conservation, 29, 30, 104
conserving, 29
constituents, 136
construction, 9, 41, 44, 62, 110
consumers, 2, 5, 7, 9, 15, 32, 40, 42, 69, 85, 91, 102, 103, 104, 117
consumption, 9, 29, 59, 60, 62, 87, 88, 93, 100, 101, 103, 112, 117, 122, 129, 131, 137, 141
convention, 49, 62
coordination, 94
copper, 80
cost, 5, 7, 8, 10, 14, 19, 27, 31, 32, 36, 40, 41, 43, 48, 49, 51, 61, 62, 67, 85, 86, 87, 91, 92, 102, 104, 107, 108, 109, 110, 123, 140
covering, 10, 24, 40, 141
crude oil, 6
currency, 8
customers, 93, 103, 104
CVD, 21

D

data collection, 78
database, 118, 133, 141
deficit, 102
Department of Commerce, 21, 46, 53, 62
Department of Energy, 89, 108, 110, 136, 139
Department of the Treasury, 110
depth, 25
developed countries, 92, 116, 125, 126, 127, 128, 129, 130, 141
developing countries, 30, 49, 62, 98, 115, 127, 128, 129, 130, 131, 141, 142
direct payment, 32

discrimination, 29, 30
displacement, 2, 3, 40, 60
distribution, 67, 68, 95, 101, 102, 132, 133
domestic demand, 2, 85, 113
domestic economy, 113, 114
domestic emissions cap, viii, 65, 70
domestic industry, 33, 68, 92, 116
domestic markets, 107
domestic policy, 113, 137
drawing, 137
dumping, 22

E

economic activity, 6, 114, 118, 121
economic concerns, viii, 65, 70
economic losses, 102
economic policy, 141
economic theory, 104
economics, 111
electric arc furnaces, 89
electricity, 6, 7, 9, 10, 13, 17, 22, 23, 34, 43, 44, 46, 51, 53, 54, 67, 77, 85, 87, 88, 91, 100, 101, 102, 103, 104, 105, 106, 110, 116, 117, 122, 129, 132, 137, 138, 140, 141, 142
emitters, viii, 45, 66, 70
employees, 5, 18, 31, 32, 60, 84
employment, 3, 5, 8, 15, 16, 17, 18, 19, 27, 31, 32, 33, 35, 43, 55, 66, 73, 82, 84, 112
employment levels, 32, 33
energy consumption, 110
energy efficiency, 68, 82, 86, 91, 108, 109, 110, 111, 135, 138
energy expenditure, 66, 71, 72, 77, 122, 137
Energy Independence and Security Act, 68, 115
energy prices, 15, 18, 42, 52, 56, 59, 60, 111, 127
energy recovery, 110
enforcement, 22, 29
environment, 92
environmental concerns, viii, 34, 65, 70
Environmental Protection Agency (EPA), v, 34, 35, 44, 46, 53, 55, 62, 63, 65, 67, 73, 77, 78, 80, 82, 83, 84, 86, 87, 88, 89, 103, 107, 112, 113, 115, 116, 117, 132, 133, 136, 137, 138, 140, 141
equilibrium, 55, 113, 133
equipment, 10, 45, 47, 50, 61, 109, 110, 112, 141
EU Emissions Trading Scheme (EU ETS), 98, 99, 103, 125, 133, 139, 141
Europe, 3, 15, 42, 52, 67, 98, 139
European Union (EU), 27, 36, 49, 58, 59, 60, 61, 63, 67, 70, 93, 95, 98, 99, 103, 116, 125, 132, 133, 139, 141
evidence, 15, 42, 56, 59, 60, 100
exchange rate, 4
exercise, 113, 117, 118
exhaustible natural resources, 29
expenditures, 71
expertise, 136
export market, 91
export subsidies, 1, 3, 5
exporters, 19
exports, 2, 4, 8, 10, 13, 16, 17, 18, 19, 21, 27, 33, 41, 47, 48, 49, 54, 77, 78, 85, 91, 93, 95, 99, 127, 137
exposure, 51, 68, 82

F

factories, 31, 47
families, 69
feedstock, 86, 109, 138, 140
financial, 49, 62, 108
financial resources, 49, 62
financial support, 108
fisheries, 16, 45, 47, 50
flexibility, 69
flights, 62
food, 45
foreign emissions, vii, 2, 8, 9, 30, 36, 39, 40
formula, 100, 104
fuel cell, 109, 110
fuel consumption, 34
fuel prices, 9, 42, 58, 122, 123, 124
funding, 110
funds, 108, 110

G

General Agreement on Tariffs and Trade (GATT), vii, 2, 5, 24, 25, 26, 28, 29, 33, 36, 37
global economy, 113, 127, 128
global trade, 133
global warming, 138
governments, 22
grants, 108, 110
greenhouse, viii, 1, 2, 3, 4, 6, 7, 8, 10, 14, 16, 18, 20, 23, 28, 34, 40, 44, 47, 65, 66, 67, 69, 70, 71, 73, 77, 78, 82, 87, 88, 93, 98, 101, 106, 107, 108, 109, 111, 115, 116, 125, 127, 135, 137, 138
greenhouse gas emissions, 47, 66, 71, 82, 101, 106, 108, 116, 125, 135, 137, 138
gross domestic product, 71

H

health, 29, 69
height, 129
historical data, 111
house, 10, 61, 69
House of Representatives, 69
human, 29
hybrid, 109, 110

I

IEA, 139
import penetration, 93
import tariffs, vii, 1, 3, 5, 26
imported products, 43
imports, 2, 3, 4, 5, 8, 10, 13, 16, 17, 18, 19, 20, 21, 22, 23, 24, 26, 27, 28, 30, 33, 36, 37, 41, 47, 48, 49, 54, 59, 67, 68, 77, 78, 85, 91, 93, 94, 95, 98, 99, 106, 107, 124, 125, 126, 127, 128, 129, 130, 131, 137, 141, 142
improvements, 68, 86, 89, 110
income, 35, 59, 102
income tax, 35
independence, 115
indirect effect, 86, 122, 123
industrial sectors, viii, 66, 67, 82, 98, 101, 107, 110, 113, 117, 135, 139, 140
industrialized countries, 3, 14, 23, 41, 58, 62
inflation, 90
information technology, 110
injury, 33
integration, 36
international competition, 9, 41, 45, 50, 92
international competitiveness, viii, 17, 32, 35, 65, 67, 70, 71, 93, 99, 100, 101, 108, 111, 112, 113, 116, 118, 125, 127, 131
International Energy Agency (IEA), 139
international investment, 56
international trade, 29, 30, 34, 37, 55, 56, 83, 113, 115, 136
investments, 80, 82, 92, 109, 110, 114, 141
iron, viii, 14, 39, 59, 67, 80, 82, 89, 98, 111, 117, 118, 125, 127
irrigation, 110
issues, 6, 25, 27, 29, 30, 34, 37, 68, 69, 107, 111, 133, 141

J

Japan, 95
job training, 109, 110

K

Kyoto Protocol, 52, 55, 58, 62, 63, 116

L

labor force, 15
labor market, 93
law enforcement, 21
laws, 21, 36
LDCs, 67, 102, 103, 104, 116, 124, 125, 126, 127, 129, 130, 131, 140, 141
lead, 2, 10, 18, 31, 47, 57, 67, 85, 86, 101, 112, 118, 122, 127, 132, 137
leadership, 136

leakage, vii, viii, 2, 3, 4, 5, 9, 10, 13, 14, 15, 18, 19, 30, 31, 34, 36, 39, 40, 41, 42, 43, 44, 45, 47, 49, 50, 52, 53, 54, 55, 56, 57, 58, 59, 60, 61, 62, 65, 66, 67, 68, 70, 71, 85, 91, 93, 98, 101, 102, 111, 112, 118, 124, 127, 128, 129, 130, 131, 132, 137,139, 141, 142
legislation, viii, 65, 66, 68, 69, 73, 114, 133, 136
letters of credit, 108
light, 10, 70, 118, 122, 127, 128
lithium, 110
lithium ion batteries, 110
loan guarantees, 108, 110
loans, 108
lower prices, 4, 10, 19

M

machinery, 47, 50
magnitude, 15, 40, 60, 62, 115
majority, 66, 68, 71, 83, 84, 86, 116, 121, 124, 130, 138, 141
management, 89
manufactured goods, 60
manufacturing, viii, 10, 16, 41, 44, 45, 47, 51, 56, 60, 62, 65, 66, 68, 69, 70, 71, 72, 73, 74, 80, 82, 83, 84, 85, 87, 92, 93, 98, 100, 109, 110, 111, 112, 115, 127, 131, 135, 137, 138
marginal product, 122
market share, 14, 16, 19, 37, 42, 47, 102
marketplace, 91
materials, 82, 89, 108, 109
metals, 3, 4, 6, 14, 16, 23, 41, 45, 48, 50, 52, 56, 66, 67, 73, 112, 117, 118, 122, 125
methodology, 77, 78
Mexico, 47, 93
microturbines, 109, 110
missions, 6, 10, 20, 41
Missouri, viii, 66
models, 52, 56, 59, 68, 112, 114, 115, 132, 136

N

natural gas, 10, 59, 86, 87, 89, 102, 103, 123, 138, 140
negative effects, 32
Netherlands, 35, 63
New Zealand, 49, 95
nitrogen, 98
nitrous oxide, 14, 47
North America, 73, 94, 133

O

OECD, 35, 139
officials, 136
oil, 9, 58, 59, 69, 86, 98, 127, 137
opportunities, 89, 131, 140, 141
organic chemicals, 82
oxygen, 89

P

participants, 116
payroll, 35
petroleum, viii, 9, 10, 14, 16, 19, 39, 40, 41, 45, 46, 48, 50, 52, 56, 58, 80, 87
physical structure, 92
plants, 10, 61, 92, 125
plastics, 56, 67, 112, 117
playing, 30
policy, viii, 1, 2, 3, 4, 5, 7, 8, 9, 10, 14, 15, 18, 19, 20, 21, 22, 24, 25, 27, 28, 29, 30, 31, 32, 33, 40, 41, 53, 58, 60, 66, 68, 69, 82, 85, 89, 107, 111, 112, 113, 114, 115, 116, 117, 118, 122, 127, 129, 130, 135, 137, 141
policymakers, 31, 36, 60
pollutants, 69
polymers, 108
pools, 19
port of export, 48
prejudice, 33, 37
president, 58, 69, 101, 134
price changes, 57

primary products, 4, 23
producers, vii, 1, 7, 8, 9, 10, 15, 19, 21, 22, 33, 34, 40, 43, 47, 50, 53, 55, 58, 69, 93, 98, 106
production costs, 2, 10, 40, 67, 85, 86, 87, 89, 91, 92, 93, 101, 102, 103, 118, 122, 123, 124, 127, 140
profit, 33, 91
profit margin, 91
profitability, 91, 139
project, 52, 55, 59, 87, 111, 112, 114
public domain, 137
publishing, 138
pulp, 67, 98, 117, 118

Q

quality standards, 93

R

ratepayers, 102, 104
ratification, 58
raw materials, 86, 89, 92, 122, 124
reality, 69, 125
reasoning, 26
recession, 106, 141
recovery, 89
regulations, viii, 24, 30, 37, 65, 70, 93, 106, 107, 115, 116, 127, 133, 137
Renewable Electricity Standard, 86
renewable energy, 86, 109, 110, 111
renewable fuel, 6, 10, 55
requirements, 10, 55, 135
researchers, 141
residential, 44
resolution, 34
response, viii, 57, 66, 92, 114, 127, 128, 129
responsiveness, 60
restrictions, 26, 29, 33
retail, 102, 105
retaliatory action, 107
revenue, 27, 32, 91
rights, 6
routes, 34
rubber, 56, 67, 117
Russia, 55, 116

S

Saudi Arabia, 9
scarcity, 136
scope, viii, 16, 66, 83, 132
Secretary of Commerce, 109, 137
Secretary of the Treasury, 137
senate, 35, 58, 62, 134
sensitivity, 42, 56, 57
September 11, viii, 66, 134
services, 34
simulation, 111
software, 6
spillovers, 93
stakeholder groups, 137
stakeholders, 69, 131, 134, 135, 136, 137
state, 27, 89, 101, 104, 128, 133
statistics, 77, 139
steel, viii, 4, 14, 23, 39, 59, 66, 67, 73, 78, 80, 82, 86, 89, 92, 98, 100, 108, 109, 111, 117, 118, 122, 124, 125, 127, 140
stockholders, 31, 32
storage, 109, 138
structure, 35, 114
subsidies, 5, 18, 22, 33, 35, 36, 43, 62
substitution, 10
suppliers, 21, 51
surplus, 57
susceptibility, 92

T

target, 58, 98, 102
tariff, 18, 26, 28
tax credits, 31, 43, 109, 110
taxes, 27, 28, 59
taxpayers, 27
teams, 3, 14, 41, 58
technological change, 114
technological progress, 114

technology, 10, 82, 86, 89, 102, 108, 109, 110
time frame, 86
Title V, 36
total costs, 51
total energy, 89, 106
trade policy, 2, 141
trade-off, 32
trading partners, 15, 49, 55, 70, 93, 107, 127, 135
training, 109, 111
transactions, 108
transition assistance, vii, 1, 5, 15, 31, 32, 33, 34, 43, 55, 62
transmission, 110
transportation, 9, 41, 44, 45, 48, 50, 92
Treasury, 136
treatment, 25, 27, 28, 37
Turkey, 62
turnover, 141

U

U.S. airline industry, 34
U.S. economy, 112, 115
U.S. policy, 28, 30, 128
UNFCCC, 62
United Kingdom, 59
United States, viii, 3, 5, 9, 10, 14, 15, 18, 20, 23, 24, 26, 29, 30, 33, 34, 37, 40, 41, 42, 44, 45, 47, 49, 50, 52, 57, 58, 62, 65, 66, 70, 73, 85, 92, 93, 105, 108, 111, 116, 125, 126, 127, 128, 129, 130, 131, 133, 134, 135, 138, 141, 142
updating, 100

V

variables, 60
vehicles, 10, 109
voiding, 28

W

wages, 15
Washington, 36, 134
waste, 89, 110
waste heat, 89
White House, viii, 66, 134
wind power, 10
wood, 52
wood products, 52
workers, 3, 8, 15, 16, 32, 82, 111
workforce, 16
World Bank, 58, 63
World Trade Organization (WTO), vii, 2, 5, 22, 23, 24, 25, 28, 29, 30, 31, 33, 34, 36, 37

Y

yield, 6, 22